GRADE 5

GEORGIA

SCIENCE

A CLOSER LOOK

BUILDING SKILLS

Reading and Writing Workbook

Macmillan/McGraw-Hill

Macmillan/McGraw-Hill

Contents

LIFE SCIENCE

Contents

EARTH SCIENCE

Contents

PHYSICAL SCIENCE

Name ______________________ Date ____________

Cells and Kingdoms

Complete the concept map by filling in answers where blanks appear.

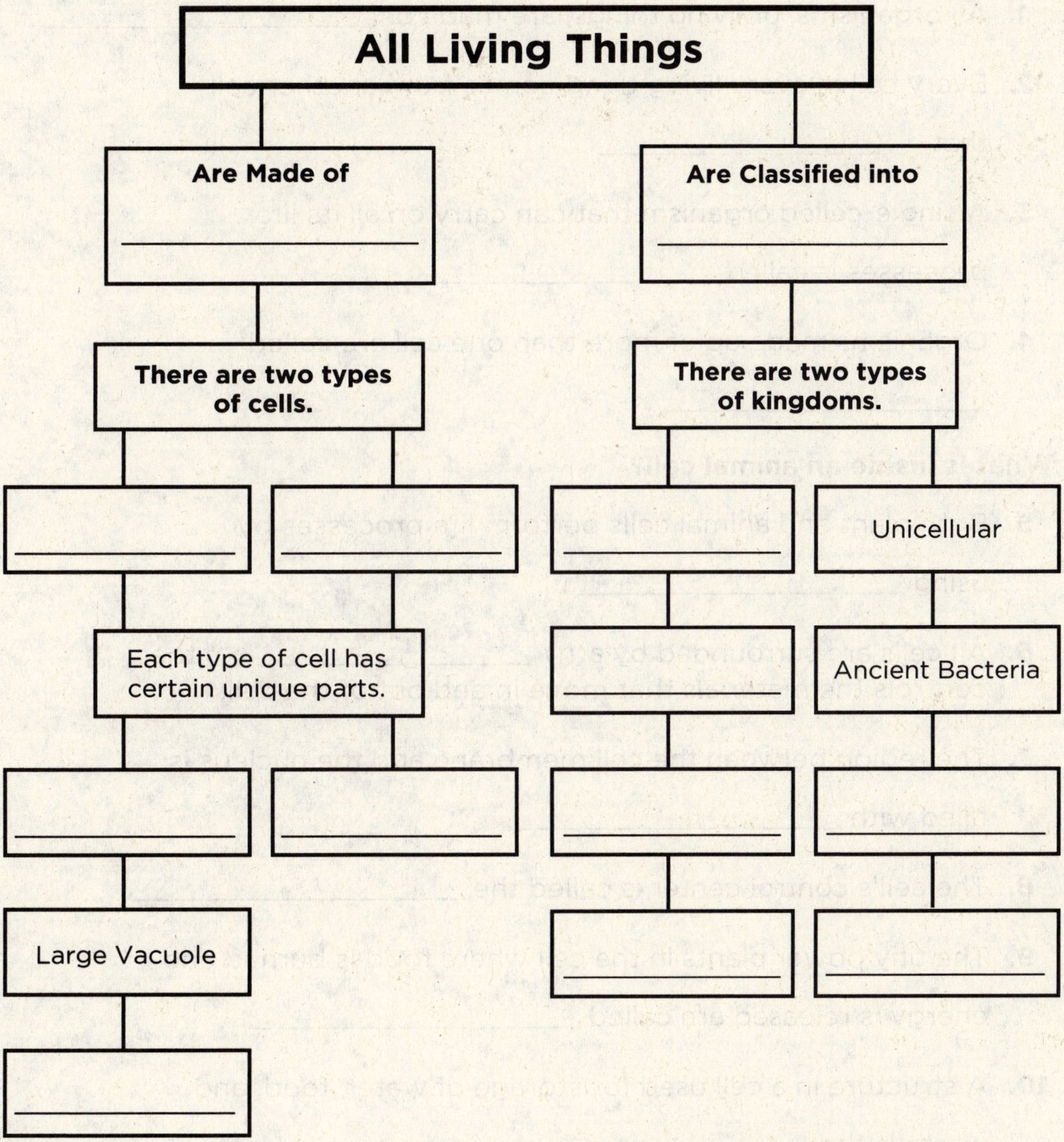

LESSON
Outline

Name ______________________ Date ____________

Cells

Use your textbook to help you fill in the blanks.

What are cells?

1. All organisms, or living things, are made of ________________.
2. Every cell in every living thing comes from another cell that ________________.
3. A single-celled organism that can carry on all its life processes is called ________________.
4. Organisms made up of more than one cell are called ________________.

What is inside an animal cell?

5. Both plant and animal cells perform life processes by using ________________.
6. All cells are surrounded by a(n) ____________ that controls the materials that move in and out of the cell.
7. The region between the cell membrane and the nucleus is filled with ________________.
8. The cell's control center is called the ________________.
9. The tiny power plants in the cell where food is burned and energy is released are called ________________.
10. A structure in a cell used for storage of water, food, and waste is the ________________.

Name ______________________ Date ____________

LESSON
Outline

What is inside a plant cell?

11. Plant cells have a(n) ______________ ; a rigid structure that serves as an outer covering.

12. A green structure, called a(n) ______________ , uses the energy from the Sun to produce food for the plant.

What is inside unicellular organisms?

13. Scientists group cells into two categories called prokaryotic and ______________ cells.

14. ______________ are cells that have a nucleus.

How are cells organized?

15. Cells working together at the same job form a(n) ______________ .

16. Groups of tissues working together form organs, and groups of organs working together form ______________ .

Critical Thinking

17. Compare and contrast the cells of plants, animals, and unicellular organisms.

____

____

____

____

____

____

____

____

LESSON
Vocabulary

Name ______________________ Date ____________

Cells

Read each clue and fill in the crossword puzzle.

cell	multicellular	organism	tissue
chlorophyll	organ	system	

Across

2. The smallest unit of a living thing that can carry out the basic processes of life.

3. Similar cells working together at the same job.

5. A green chemical that absorbs sunlight.

7. Another name for a living thing.

Down

1. Organisms that contain many different types of cells.

4. Organs that work together to do a certain job.

6. A group of tissues working together to perform a specific job.

Name ______________________ Date ____________

Cells

Fill in the blanks.

cell membrane	cytoplasm	nucleus
cell wall	mitochondria	organelles
cells	multicellular	unicellular

When we talk about living things, we can break them down into smaller things. All living things are made up of units called ______________ . Some organisms are ______________ ; that is, they consist of only one cell. More complex organisms, including plants and animals, are called ______________ organisms.

All cells are surrounded by a(n) ______________ that controls what moves into and out of the cell. The insides of cells are filled with a gel like fluid called ______________ . Within this liquid are the cell ______________ . Both plant and animal cells, as well as many unicellular organisms, contain a(n) ______________ and ______________ , which supply energy for the cell. Plant cells have a(n) ______________ , one large central vacuole, and chloroplasts. Chloroplasts contain chlorophyll which uses energy from sunlight to produce food for the plant.

LESSON **Outline**

Name ______________________ Date ____________

Classifying Life

Use your textbook to help you fill in the blanks.

How are organisms classified?

1. Scientists sort, or ____________, living things into groups according to shared traits.

2. All organisms are divided into six major groups called ____________.

3. The six subgroups used to classify organisms within kingdoms are ____________, class, ____________, family, genus, and ____________.

4. The scientific name of an organism consists of its ____________.

What are animals?

5. The two kingdoms that include only multicellular organisms are ____________.

6. Plants can make their own food, but ____________ obtain energy from other organisms.

7. The two major groups of animals are ____________ and ____________.

What are plants and fungi?

8. Although they are from two different kingdoms, plants and fungi always have ____________.

Name ______________________________ Date ____________

LESSON
Outline

9. The two major groups into which plants are organized are

______________________ and ______________________ .

10. Unlike plants, ______________________ get food by breaking down dead organisms.

11. A fungus that makes bread rise is called ______________________ .

What are bacteria and protists?

12. Bacteria are unicellular organisms with no ______________________ or mitochondria.

13. The two kingdoms used to classify bacteria are the

______________________ and ______________________ .

14. Protists can be ______________________ or multicellular.

15. Unlike bacteria, protists have large cells, a nucleus, and

bound ______________________ .

What are viruses?

16. Viruses are not classified as living organisms because they

carry out no life processes except ______________________ .

Critical Thinking

17. What makes plants and animals different from bacteria?

____

____

____

____

____

Name ______________________ Date __________

Classifying Life

Match the correct letter with the description.

a. classification	**c.** kingdom	**e.** species	**g.** vertebrate
b. invertebrate	**d.** nonvascular	**f.** vascular	**h.** virus

1. ________ contains only closely related organisms
2. ________ the broadest group into which an organism can be classified
3. ________ helps scientists identify, study, group, and name organisms
4. ________ means "contains tubes or vessels"
5. ________ animal without a backbone
6. ________ type of plant that tends to be small and close to the ground
7. ________ animal with a backbone, nervous system, and brain
8. ________ organism that carries out only one life process, reproduction

Name ______________________________ Date ____________

Classifying Life

Fill in the blanks.

classify	kingdom	species
genus	scientific name	unicellular

Living things are often similar to each other. Scientists ________________ living things by similarity into smaller and more specific groups. The largest group into which an organism is classified is its ________________. The six kingdoms include plants, animals, fungi, protists, "true" bacteria, and "ancient" bacteria.

Each kingdom is divided into progressively smaller groups, as follows: phylum, class, order, family, genus, and species. The narrowest group into which an organism can be classified is a(n) ________________. Two words make up the ________________ of a living thing. The first is the ________________ name. The second is the species name.

The "true" bacteria and the "ancient" bacteria are ________________ organisms with no nucleus or mitochondria. "Ancient" bacteria usually live in very harsh environments in which other organisms could not survive.

Name ______________________ Date ____________

Meet Angelique Corthals

Getting Ideas

Underline the sentence or sentences in each paragraph that state the main idea of the paragraph.

How can you fit thousands of organisms into one small room? Angelique Corthals knows. She's a scientist at the American Museum of Natural History, and she's been busy preserving tissue samples of many different organisms from around the globe, including samples from species that have become extinct or died out. Angelique is an archaeologist. That's a scientist who studies the past.

Angelique works in the museum's frozen tissues lab. She specializes in the preservation of the information in cells. From bacteria to insects to mammals, she collects, preserves, and organizes the cells of all sorts of living things. Angelique stores the organisms' cells and freezes them in small plastic tubes the size of your finger. Just as food stays fresh in the freezer, freezing cells prevents them from spoiling or decomposing. The tubes are stored in large tanks containing liquid nitrogen. At -150°C (-238°F), this liquid is so cold that all of the cells' biological processes stop.

By using this freezing process, the cells can be preserved for many years. When a scientist needs to study an organism, she can request a cell sample from the lab. Whether it's from a small fly or a large humpback whale, each cell contains information about the whole organism. Scientists can use this information to learn how different organisms are related. They can also use this information to learn about living things that have already become extinct and to understand why they died out.

Name ______________________________ Date ____________

Write About It

Summarize Make a chart that tells the steps for preserving cells. Use your chart to write a summary of the process Angelique uses to freeze cells from organisms.

Sequence Make a chart that tells the steps for preserving cells. Use the blank boxes below.

First

↓

Next

↓

Last

Summarize In a paragraph, summarize the process that Corthals uses to freeze cells.

____

____

____

____

LESSON **Outline**

Name ______________________ Date ____________

Plants

Use your textbook to help you fill in the blanks.

How are plants classified?

1. Small plants such as mosses which survive without a transport system, are called ______________________ .
2. Plants that have a system of hollowed-out tubes to transport water and nutrients are called ______________________ .
3. A seed plant that does not produce flowers or fruits is called a(n) ______________________ .
4. A seed plant that produces flowers and some kind of fruit is called a(n) ______________________ .

What are roots?

5. Roots absorb minerals and water, store food, and ______________________ .
6. Root hairs absorb water and minerals, and ______________________ protect root tips.
7. The epidermis is on the outside of the root; just beneath it is the ______________________ , used to store food; and at the center of the root is the ______________________ .

Name ______________________ Date ____________

LESSON
Outline

What are stems?

8. Stems have two main functions: ______________ and transport.

9. Grasses have ______________ stems that are green and bendable; trees have ______________ stems.

10. Two kinds of cells make up the transport system of plants. Cells that move water and minerals up the plant are ______________. Cells that move food from the leaves to other plant parts are ______________.

What are leaves?

11. The function of leaves is to perform ______________.

12. To perform photosynthesis, chloroplasts need ______________ from the air, water from the soil, and ______________.

13. Air enters and exits plants through ______________: pores on the underside of the leaves.

Critical Thinking

14. Why do you think some plants have woody stems and some have soft stems?

____

____

____

____

____

____

Name ______________________ Date ____________

Plants

Fill in the blank with a term from the box.

angiosperm	gymnosperm	transpiration
cambium	phloem	xylem
cellular respiration	photosynthesis	

1. A seed that does not produce a flower is called a(n) ________________ .

2. Cells that move sugars up, down, and all around a plant are called ________________ .

3. A layer in the plant stem that separates xylem and phloem is called the ________________ .

4. The break down of sugars in plant and animal cells to produce energy and carbon dioxide is called ________________ .

5. A seed that produces a flower is called a(n) ________________ .

6. Cells that transport water and minerals from roots to shoots in plants are called ________________ .

7. The process that plants use to produce their food and give off oxygen is called ________________ .

8. When water moves up the vascular tubes through stomata, ________________ occurs.

Name ______________________________ Date ____________

Plants

Fill in the blanks.

angiosperms	gymnosperms	stomata
cellular respiration	photosynthesis	sugar
chloroplasts	stems	

Seed plants can be divided into two main groups. Scientists call these groups ____________________ (flowering plants) and ____________________ (plants without flowers or fruits). Seed plants have three basic parts—leaves, roots, and ____________________ .

The function of leaves is to absorb sunlight to make sugars, a process called ____________________ . The energy of sunlight is captured by ____________________ and used to combine carbon dioxide and water. The carbon dioxide comes into the leaf through ____________________ . The sugar made during photosynthesis travels to cells all over the plant, where it is used for ____________________ . During this process, ____________________ is broken down to release energy to power the cell's functions. The by-products are carbon dioxide and water.

Name ______________________ Date __________

Classifying Animals

Use your textbook to help you fill in the blanks.

What are simple vertebrates?

1. The simplest animals are ________________ .
They are without real tissues or organs and have a(n)
________________ body plan.

2. Jellyfish and hydras are ________________ .
They possess a mouth and muscle tissue and are
________________ symmetrical.

3. Worms that have flat bodies with one body opening
and simple eyes are called ________________ .

4. Worms that have simple digestive and nervous systems
are called ________________ .

What are complex invertebrates?

5. Clams and squids are ________________ . They have
________________ symmetry, a muscular foot, a
mantle, and several specialized organs.

6. Sea stars and sea cucumbers are ________________ .
They have ________________ feet and a water
pressure system that helps them feed, breathe,
and move.

7. Crabs and insects belong to the largest animal group on
Earth, the ________________ phylum.

Name ______________________ Date ____________

LESSON
Outline

What are vertebrates?

8. There are three kinds of fish: ____________, such as lamprey and hagfish; ____________, such as sharks and skates; and ____________.

9. Frogs, toads, and salamanders are ____________.

10. Lizards, snakes, turtles, and alligators are ____________. They are ____________, which means that their body temperature is not steady.

11. Birds are designed for flying: they are warm-blooded and have ____________ and feathers that are light and strong.

What are mammals?

12. Animals that are warm-blooded and have hair are called ____________.

13. A duck-billed platypus lays eggs. It is a(n) ____________.

14. A kangaroo is a(n) ____________. It gives birth to partially developed offspring.

15. Lions, whales, and humans are ____________. Their offspring develop within the mother.

Critical Thinking

16. Compare 4 different vertebrates.

____

____

____

____

LESSON
Vocabulary

Name ______________________ Date ____________

Classifying Animals

Read each clue and fill in the blank with the correct answer.

asymmetrical	invertebrates	monotreme	radial
bilateral	marsupial	placental	vertebrates

1. ____________________ A koala is one. It gives birth to partially developed offspring.

2. ____________________ Worms have this kind of symmetry because they can be divided along only one plane.

3. ____________________ Fish, birds, amphibians, reptiles, and mammals.

4. ____________________ A whale is an example of this kind of mammal.

5. ____________________ Cnidarians have this kind of symmetry.

6. ____________________ A mammal that lays eggs.

7. ____________________ A type of body plan that has no definite shape.

8. ____________________ Sponges, coelenterates, echinoderms, mollusks, and arthropods.

Name ______________________ Date ____________

LESSON
Cloze Activity

Classifying Animals

Fill in the blanks.

amphibians	invertebrates	sponges	vertebrates
hollow	reptiles	tentacles	

The animal kingdom contains all the animals. The animal kingdom is separated into two large groups: animals with backbones (________________) and animals without backbones (________________) . These two groups are divided into smaller groups called phyla.

Vertebrates include fish, amphibians, reptiles, birds and mammals. Fish live in the water and breathe through gills. Vertebrates that spend part of their lives in water and part on land are called ________________ . Lizards, snakes, turtles, alligators, and crocodiles are ________________ . Birds are designed for flying. Their bones are ________________ and light. Mammals produce milk to feed their young.

Invertebrates include sponges, coelenterates, worms, and arthropods. The most primitive of the animal groups are called ________________ . Coelenterates have mouths surrounded by stinging ________________ . The largest of all the animal groups are called arthropods.

Name ______________________ Date __________

The Underground Life of Mole Rats

Write About It

Find out the scientific name of an animal you think is cute or ugly. Write a description of the animal. Use words and details that appeal to the senses in your description.

Getting Ideas

Choose an animal to describe. Then use the web below to brainstorm ideas. Write its scientific name in the center circle. Write details that describe it in the outer circles. You can add circles to the web if you like.

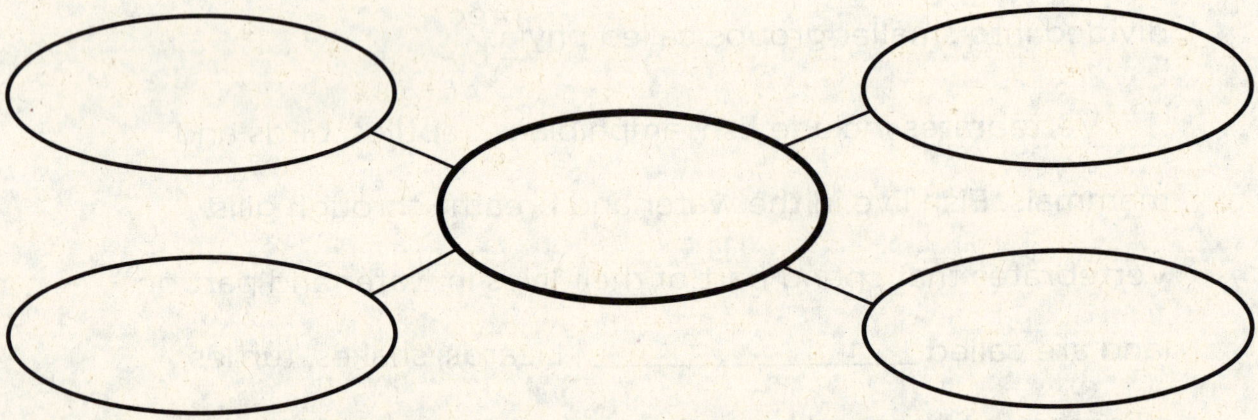

Planning and Organizing

Janine wants to describe a ferret. Here are some sentences that she wrote. Write Yes if the sentence contains words and details that create a vivid impression of the ferret. Write No if it does not.

1. ________ The dark black marks around its eyes make it look like a bandit.

2. ________ Its long, lean body curves upward as though in surprise.

3. ________ The ferret has four legs and a tail.

Drafting

Write a sentence to begin your description. Tell what animal you will describe. Make sure your sentence expresses the impression you want to create.

____

____

Now write your description. Use a separate piece of paper. Begin with the sentence you wrote above. Tell the animal's scientific name. Include descriptive words and vivid details to help readers experience the animal.

Revising and Proofreading

Here are three sentences that Janine wrote for her description. Help her improve them. Replace each italic word or words with a more descriptive or precise word. Choose a word from the box. Write it in the blank.

bushy	carnivores	nocturnal

1. ____________________ Since ferrets are *active during the night*, you will probably find them curled up and sleeping most of the day.

2. ____________________ Some ferrets have a *hairy* tail with an inky-black tip.

3. ____________________ Ferrets are *animals that eat meat*. They may eat squirrels, mice, and even prairie dogs.

Now revise and proofread your writing. Ask yourself:

- ▶ Did I create a vivid impression of the animal I chose?
- ▶ Did I use sensory words and vivid details?
- ▶ Did I correct all mistakes in grammar, spelling, punctuation, and capitalization?

LESSON **Outline**

Name ______________________ Date __________

Microorganisms

Use your textbook to help you fill in the blanks.

What is a microorganism?

1. Scientists need ______________ to study the tiny creatures called microorganisms.

2. In a compound microscope, the ______________ is placed near the object being studied. The ______________ is near the scientist's eye.

3. These microscopes can ______________ objects as much as 2,000 times.

4. A(n) ______________ microscope is even more powerful. It can magnify objects as much as 300,000 times and show objects in ______________ dimensions.

How are microorganisms beneficial?

5. Microorganisms provide ______________ to other organisms.

6. Decomposers break down ______________ into simpler substances.

7. Bacteria change ______________ into a form that plants use.

8. Cottage cheese is made by ______________ changing milk.

9. Scientists often use bacteria as ______________ for research because they are simple organisms.

Name ______________________ Date ____________

LESSON **Outline**

How are microorganisms harmful?

10. Some bacteria cause ________________ by attacking cells in the lungs.

11. Doctors test for the presence of ________________ by swabbing the back of the throat to find evidence of the bacteria that can cause bad sore throats.

12. Bacteria that live in uncooked foods can cause illnesses by attacking the ________________ .

13. Microorganisms that cause harm to other organisms that they live on are called ________________ .

14. The disease ________________ is caused by a parasitic protist that lives in mosquitoes and humans.

How do we protect ourselves from harmful microorganisms?

15. Antibiotics are used to treat bacterial infections. The first antibiotic used was ________________ .

16. Heating food by cooking and by ________________ destroys the bacteria that may be present in the food.

17. One of the best ways to stop the spread of infections is ________________ .

Critical Thinking

18. What advantages does the scanning electron microscope have over a compound microscope?

____

____

____

Name ____________________ Date __________

Microorganisms

Who am I? What am I?

Choose a word from the word box below that answers each question.

a. decomposer	**d.** parasite	**g.** unicellular
b. microorganism	**e.** protist	
c. microscope	**f.** spore	

1. ________ I am produced when male and female cells fuse together, and then I carry genetic information within a protective coating. What am I?

2. ________ I am an instrument used to look at microscopic things. What am I?

3. ________ I am an organism that is not visible to the unaided eye. What am I?

4. ________ I live on other organisms and cause them harm. What am I?

5. ________ I am another way to describe a single-celled organism. What am I?

6. ________ I am a type of organism that breaks down dead material. What am I?

7. ________ I am not a fungus or a bacteria, but I am a microorganism. What am I?

Name ______________________ Date ____________

LESSON
Cloze Activity

Microorganisms

Fill in the blanks.

Antibiotics	oxygen	strep throat
decompose	parasites	
food	protists	

Microorganisms are an important part of Earth's ecosystems. They provide most of the ______________ in the atmosphere. Bacteria are one kind of microorganism. Some bacteria are beneficial to the environment and help ______________ dead material. Many microorganisms can also produce their own ______________. They provide a source of food for larger organisms.

There are some microorganisms that are harmful to other forms of life. These microorganisms are called ______________. They cause harm to humans, plants, and animals. Three common harmful microorganisms are fungi, ______________, and bacteria. Bacteria cause many infectious diseases, such as pneumonia and ______________.

Doctors and scientists have discovered ways to treat infections. ______________ were discovered in 1928 by Alexander Fleming as a way to treat bacterial infections. One of the best ways to stop infections from spreading is hand-washing.

Name ______________________ Date ____________

Meet Maria Pia Di Bonaventura

Read the Reading in Science feature in your textbook. As you read, look for the topic sentence in each paragraph. Then look for supporting details about each topic sentence.

Main Idea

Use the graphic organizer to record the topic sentence and supporting details for each paragraph of the article.

Topic Sentence	Supporting Details

Name ______________________ Date ____________

Write About It

Main Idea and Details

1. Why do museums ask for help from scientists like Maria Pia Di Bonaventura?
2. How does Maria Pia's work help protect works of art and other artifacts?

Planning and Organizing

Write the problem that museums sometimes have with their artifacts and works of art.

____

____

____

____

Write a one-sentence summary of Maria Pia's job.

____

____

____

Drafting

Now explain how Maria Pia's area of expertise can help museums protect their works of art.

____

____

____

____

____

Name ____________________ Date __________

Cells and Kingdoms

Choose the letter of the best answer.

1. The flexible wrapping that surrounds all cells is called the

a. cytoplasm.

b. cell wall.

c. cell membrane.

d. vacuole.

2. Which of these structures is found in a plant cell, but not in an animal cell?

a. central vacuole

b. mitochondria

c. nucleus

d. organelles

3. Which of the following is in the correct order, from simplest to most complex?

a. cells, tissues, organs, organism, organ systems

b. cells, tissues, organs, organ systems, organism

c. cells, organs, tissues, organ systems, organism

d. cells, tissues, organ systems, organs, organism

4. Which phylum has the most species?

a. echinodermata

b. molluska

c. chordata

d. arthropoda

5. In which kingdom do all of the members obtain energy by breaking down dead organisms?

a. plants

b. animals

c. fungi

d. protists

6. Which life process do viruses carry out?

a. reproduction

b. movement

c. growth

d. use of energy

7. Which structure is found in vascular plants, but NOT in nonvascular plants?

a. flower

b. xylem

c. leaf

d. seed

Choose the letter of the best answer.

8. A fir tree is an example of a(n)

a. nonvascular plant.

b. angiosperm.

c. seedless plant.

d. gymnosperm.

9. A carrot root is an example of a(n)

a. taproot.

b. prop root.

c. fibrous root.

d. aerial root.

10. Which of the following moves sugars that are made in the plant's leaves?

a. xylem

b. phloem

c. cambium

d. bark

11. Which group includes the simplest animals?

a. worms

b. mollusks

c. sponges

d. cnidarians

12. Which animals have radial symmetry?

a. sponges

b. worms

c. mollusks

d. cnidarians

13. Which vertebrates have gills when they hatch, and lungs when they are adults?

a. amphibians

b. reptiles

c. fish

d. birds

Name ______________________ Date ____________

Parents and Offspring

Complete the concept map with information you have learned about different types of reproduction. Some answers have been written for you.

All Living Things Reproduce

Types of Reproduction	Organisms that use this type of reproduction	Does this type of reproduction enhance genetic variation?	Disadvantages or Advantage to this type of reproduction
____________	____________	____________	____________ ____________ ____________
____________	bacteria	____________	____________ ____________ ____________ ____________

Name ______________________ Date ____________

LESSON **Outline**

Reproduction

Use your textbook to help you fill in the blanks.

What are sexual and asexual reproduction?

1. Survival of a(n) ____________________ depends on its ability to produce offspring.
2. Every organism comes from a parent through the process of ____________________ .
3. The transfer of ____________________ from parents to their offspring is known as reproduction.
4. Genetic material contains the information that controls an organism's ____________________ .
5. The production of a new organism from two parents is called ____________________ reproduction.
6. When an egg cell joins with a sperm cell, ____________________ occurs.
7. A fertilized egg develops into an individual with traits from each ____________________ .
8. The production of a new organism from a single parent is called ____________________ reproduction.
9. Most bacteria and unicellular protists reproduce by making a copy of their genetic material and ____________________ .
10. Cnidarians, sponges, and some fungi can reproduce through ____________________ .

Name ______________________ Date ____________

11. The eggs of insects, fish, frogs, and lizards sometimes develop into new animals without being ____________ .

12. New plants can grow from leaves, roots, or stems. This type of asexual reproduction is called ____________ .

13. Strawberry plants and ferns can reproduce asexually by forming ____________ .

How do sexual and asexual reproduction compare?

14. An organism that reproduces asexually does not have to find a(n) ____________ .

15. Organisms that reproduce asexually tend to be well suited to their ____________ .

16. A major advantage of sexual reproduction is that it promotes ____________ in a species.

Critical Thinking

17. Why is sexual reproduction better than asexual reproduction for ensuring the survival of a species in a changing environment?

Name ______________________ Date __________

Reproduction

Read each clue. Write the answer in the blanks using the words below. Then fill in the crossword puzzle.

asexual	runners	splitting	variation
budding	sexual	trait	vegetative

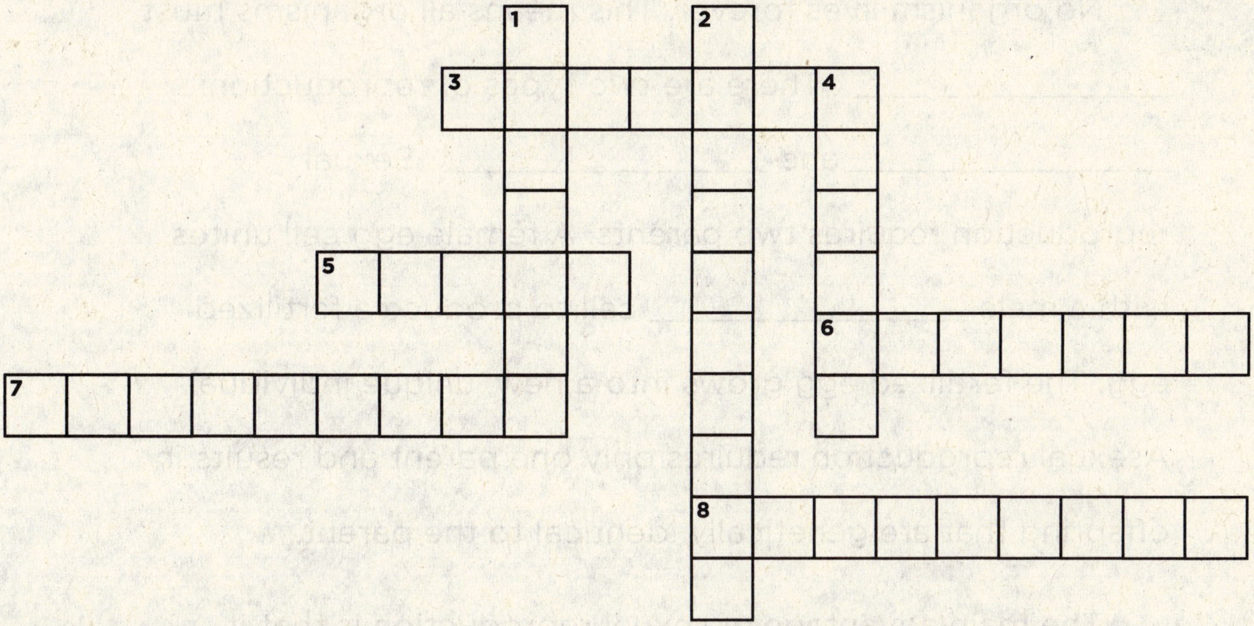

Across

3. Plant stems that run along the ground and sprout as new plants.
5. Any characteristic of a living thing.
6. The production of a new organism from one parent is called ______________ reproduction.
7. Bacteria reproduce by ______________.
8. Sexual reproduction gives rise to ______________ in a species.

Down

1. A bud growing from a fungus to become a new individual.
2. A new plant growing from a leaf is ______________ propagation.
4. A new organism from two parents is ______________ reproduction.

LESSON
Cloze Activity

Name ______________________ Date __________

Reproduction

Fill in the blanks.

asexual	mate	sexual	variety
fission	reproduce	sperm	

No organism lives forever. This means all organisms must ______________ . There are two types of reproduction: ______________ and ______________ . Sexual reproduction requires two parents. A female egg cell unites with a male ______________ cell to produce a fertilized egg. The fertilized egg grows into a new, unique individual. Asexual reproduction requires only one parent and results in offspring that are genetically identical to the parent.

The main advantage of sexual reproduction is that it promotes ______________ within a species. An advantage of asexual reproduction is that it does not require finding a(n) ______________ . There are several methods of asexual reproduction. Simple, one-celled organisms, like bacteria and protists, undergo ______________ . Animals such as cnidarians, sponges, and fungi undergo a process called budding.

Name ______________________ Date __________

How Do Sea Stars Regenerate?

Write About It

The article you just read explained how the sea star can produce offspring using regeneration. Choose another animal that can reproduce without two parents. Write an explanation of how this process takes place.

Getting Ideas

Choose an animal to write about. Think about how it reproduces without parents. Write the steps below.

First

↓

Next

↓

Last

Planning and Organizing

Xavier wants to explain how flat worms reproduce. Here are three sentences he wrote. Put them in order.

________ Finally, each half grows into a separate flat worm.

________ First, the flat worm divides in two.

________ Stem cells turn into the types of cells needed to reproduce the missing part.

Name ______________________ Date __________

Drafting

Write a sentence to begin your explanation. Name the animal you are writing about. Tell your main idea about how this animal reproduces. This is your topic sentence.

____

____

____

Now write your explanation. Use a separate piece of paper. Begin with your topic sentence. Explain how the animal reproduces. Write the steps in time order.

Revising and Proofreading

Here are some sentences Xavier wrote. Combine each pair. Use the time order word in parentheses. Write the new sentence on the line.

1. The stem cells multiple. They turn into specialized cells. (before)

____

____

2. A message is sent out to specialized cells. The cells near the wound cover it. (after)

____

____

Now revise and proofread your writing. Ask yourself:

- ▶ Did I explain how the animal can reproduce without parents?
- ▶ Did I include time order words?
- ▶ Did I correct all mistakes?

Name ______________________ Date ____________

LESSON **Outline**

Traits and Heredity

Use your textbook to help you fill in the blanks.

What is heredity?

1. The passing of traits from one generation to the next is called ____________________.

2. Traits that offspring receive from their parents are ____________________ traits.

3. A way of acting or behaving with which an animal is born is called a(n) ____________________.

4. A behavior that develops during an animal's lifetime is a(n) ____________________ behavior.

5. When ducks hatch, they learn to recognize and follow their mother, a behavior called ____________________.

How are traits inherited?

6. Mendel discovered that each inherited trait is controlled by ____________________, one from each parent.

7. Today scientists refer to Mendel's factors as ____________________.

8. Genes are found in the nucleus of the cell. They are stored on ____________________.

9. A trait that masks another trait is called a(n) ____________________ trait.

10. A trait that is masked is called a ____________________ trait.

Name ______________________ Date ____________

11. In pea plants, purple flowers are a dominant trait and white flowers are a recessive trait. The purple trait is represented by ____________________ and the white trait by p.

How do we trace inherited genes?

12. A chart used to trace the history of traits in a family is called a(n) ____________________ .

13. On a pedigree chart, horizontal lines connect parents and vertical lines connect parents to ____________________ .

14. Males are represented by squares, and ____________________ are represented by circles.

15. Shaded shapes represent individuals with a particular ____________________ , and unshaded shapes represent individuals without that trait.

16. Dimples are a dominant trait, represented by the letter D. A child who is a carrier of the recessive trait is represented by ____________________ .

Critical Thinking

17. Both a father and mother have dimples. Their son has dimples, but their daughter does not. Which genes, DD, Dd, or dd, does each family member have?

____

____

____

____

____

Name ______________________ Date ____________

Traits and Heredity

Match the correct letter with the description.

a. carrier	**d.** heredity	**g.** pedigree
b. dominant	**e.** inherited	**h.** recessive
c. gene	**f.** instinct	

1. ________ a trait that an offspring receives from its parents

2. ________ the passing down of traits from one generation to the next

3. ________ behavior that is inherited

4. ________ a trait that masks another trait

5. ________ a trait that is masked or covered by another trait

6. ________ chart used to trace the history of traits in a family

7. ________ contains the chemical instructions for an inherited trait

8. ________ individual who has inherited a gene for a trait, but does not show the trait physically

LESSON
Cloze Activity

Name ______________________ Date __________

Traits and Heredity

Fill in the blanks.

chromosomes	heredity	Gregor Mendel	sperm cell
genes	instincts	pedigree	trait

Parents pass on features of themselves to their offspring. Any notable feature of an organism is called a(n) ________________. The passing down of traits from parents to offspring is called ________________. Some traits, such as hair or eye color, are physical traits. Other inherited traits are behavioral and are called ________________. An Austrian monk, ________________, discovered how traits are inherited.

Today, Mendel's factors are called ________________. They are stored on the ________________ inside the nucleus of cells. Offspring receive one set of genes from an egg cell and the other from the ________________ that fertilized the egg cell.

Humans have an estimated 20,000 gene pairs. Some of these traits are easy to see. The history of a family trait and the way it has been inherited can be charted in a ________________. These charts can be used to study heredity patterns.

Name ______________________ Date ____________

Genetically Modified Corn

Read the Reading in Science feature from your textbook. Look for cause and effect relationships.

Cause and Effect

Fill in the Cause and Effect Chart with cause and effect relationships you find in the article.

Cause		Effect
Corn borer eats corn	→	
Bt powder sprayed on corn	→	
	→	Corn plants make Bt toxin in their own cells, so the corn plants can protect themselves.
Other living things eat Bt corn	→	

Name ______________________ Date __________

Write About It

Cause and Effect Explain how the bacterium Bt affects corn borers. Tell how genetically modified corn might cause problems for other insects and for the environment in general.

Planning and Organizing

Answer these questions in detail.

1. What does the Bt bacterium produce, and what effect does it have on corn borers?

____

____

2. What enables the Bt bacterium to make a protein that is toxic to corn borers?

____

____

3. What was transferred from the Bt bacterium to Bt corn?

____

____

4. How does Bt corn affect corn borers?

____

5. How might Bt corn affect other living things, such as Monarch butterflies?

____

____

Name ______________________ Date ____________

Parents and Offspring

Choose the letter of the best answer.

1. Which of the following organisms reproduces by using budding?

a. sponge **c.** lizard
b. cat **d.** frog

2. Which of the following plants reproduces by using runners?

a. corn plant
b. moss
c. strawberry plant
d. apple tree

3. Which of the following is an example of sexual reproduction?

a. cloning
b. budding
c. seed production
d. vegetative propagation

4. Which organism can develop from an unfertilized egg?

a. human **c.** bird
b. sheep **d.** lizard

5. Which of the following is an advantage of asexual reproduction?

a. It depends on finding another organism.
b. It promotes variety in a species.
c. It is convenient.
d. It gives rise to offspring better suited to environmental change.

6. When a sperm cell and an egg cell join together, they form

a. part of both the male and female parent.
b. a new organism.
c. part of the male parent.
d. part of the female parent.

7. The passing of traits from parents to their offspring is known as

a. cloning. **c.** instinct.
b. heredity. **d.** behavior.

Name ______________________ Date ____________

Choose the letter of the best answer.

8. An example of a behavior that is not inherited is

a. a dog playing fetch.

b. a bird building a nest.

c. a child crying.

d. a spider building a web.

9. Genes are found

a. in the vacuole of a cell.

b. on the cell membrane of a cell.

c. outside of a cell.

d. in the nucleus of a cell.

10. A carrier

a. inherits the genes for a trait and shows the trait.

b. does not inherit the genes for a trait and does not show the trait.

c. inherits the genes for a trait, but does not show trait.

d. does not inherit the genes for a trait, but does show the trait.

11. Which of the following items represents a carrier for the recessive trait?

a. DD **c.** dd

b. Dd **d.** d

12. An instinct is an example of

a. a learned behavior.

b. an inherited behavior.

c. an inherited physical trait.

d. imprinting.

13. If purple is the dominant gene for flower color, which of following items represents a white flower?

a. PP **c.** Pp

b. pp **d.** p

Name ______________________ Date ____________

Adventures in Eating

Read the Literature feature in your textbook.

Write About It

Response to Literature This article tells about different adaptations for eating. Research two more animals that have interesting adaptations. Write a report that explains how these adaptations help the animals eat. Compare these adaptations to the ones you read about in the article.

CHAPTER
Concept Map

Name ______________________ Date __________

Our Dynamic Earth

Complete the concept map by filling in answers where blanks appear.

Forces that Change Earth's Surface

- Water, Ice, Wind, and Gravity
  - ______________ → River ______________; Sandbars, ______________ islands
  - Erosion → ______________, mudslides; Cirques, ______________ valleys
  - ______________ → Breaks down ______________ of all landforms.
- Plate Movements
  - Move over hot spots → Island ______________
  - ______________ ______________ → ______________ mountains; Earthquakes
  - ______________ → Mid-ocean ______________; Earthquakes
  - Move toward each other → ______________ mountains; Earthquakes

Name ______________________ Date ____________

LESSON
Outline

Earth's Landforms

Use your textbook to help you fill in the blanks.

What are landforms?

1. A physical feature on Earth's surface is a(n) ______________.
2. The highest of Earth's physical features are ______________.
3. A low area between mountains or hills is a(n) ______________.
4. Wide, flat areas of land are called ______________.
5. A large, flat area higher than the land around it is a(n) ______________.
6. Earth's largest bodies of water are its saltwater ______________.
7. Natural streams of flowing water that empty into lakes, oceans, or other bodies of water are ______________.
8. A body of water with land all around it is a(n) ______________.

What are the features of the ocean floor?

9. The largest landform under the ocean is an ocean ______________.
10. Shallow waters cover the ______________, the gently sloping part of the ocean floor along the coast.
11. The sharp drop from the continental shelf to the continental rise is the ______________.

Name ____________________ Date __________

12. A wide, flat area covering about 40 percent of the ocean floor is the ____________.

13. The deepest areas of the ocean floor are ____________.

How do we map Earth's features?

14. Measurements taken by a(n) ____________ are used to make maps.

15. Elevations are shown with shading on a(n) ____________ map.

16. Lines are used to show elevation and steepness of slopes on a(n) ____________ map.

Where are Earth's layers?

17. The layer of air around Earth is the ____________.

18. Earth's waters make up Earth's ____________.

19. Earth is made of three main layers: the crust, the ____________, and the core.

20. The part of Earth that is home for living things is the ____________.

Critical Thinking

21. Compare the mantle and core of the Earth.

____

____

____

____

Name ______________________ Date ____________

Earth's Landforms

Match the correct word with its description, and fill in the crossword puzzle.

atmosphere	crust	landform	mantle
core	hydrosphere	lithosphere	

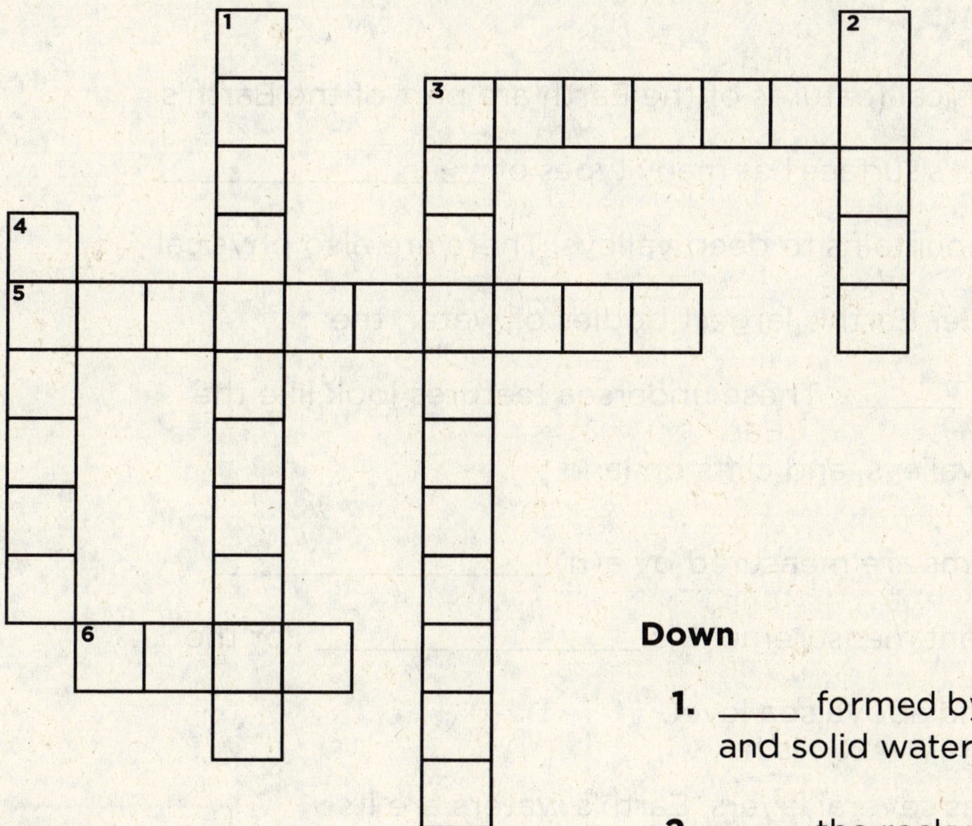

Across

3. _____ a physical feature on Earth's surface

5. _____ the layer of air that surrounds Earth

6. _____ the central part of Earth

Down

1. _____ formed by Earth's liquid and solid water

2. _____ the rocky upper layer of Earth that contains continents and ocean basins

3. _____ the crust and the top of the upper mantle form it

4. _____ the layer of Earth's interior below the crust

Name ______________________ Date ____________

Earth's Landforms

Fill in the blanks.

crust	inner core	oceans
elevation	landforms	outer core
hydrosphere	mantle	surveyor

The physical features of the Earth are part of the Earth's surface. Earth's surface has many types of ________________, from high mountains to deep valleys. There are also physical features under Earth's largest bodies of water, the ________________. These undersea features look like the mountains, valleys, and cliffs on land.

Landforms are measured by a(n) ________________. One important measurement is ________________, or the height of land above sea level.

Earth has several layers. Earth's waters are its ________________. The planet itself is divided into the surface ________________, the ________________ beneath it, and the core at the center. The crust and mantle are rock. Earth's core is made of metal. The ________________ is liquid, and the ________________ is solid. The core makes up the central part of the Earth.

Name ____________________ Date __________

LESSON **Outline**

Plate Tectonics

Use your textbook to help you fill in the blanks.

Is Earth's crust moving?

1. Geologist Alfred Wegener formulated the theory of ____________________ .

2. Wegener's theory stated that Earth's ____________________ were once joined in one landmass, but gradually pulled apart and drifted.

3. Wegener's showed that the age and composition rocks in the ____________________ on South America's east coast matched of those on Africa's west coast.

4. Scientists also discovered evidence in ____________________ that Africa and South America were once joined.

How does the movement of Earth's crust affect the ocean?

5. Scientists developed the ____________________ model to explain how the continents have moved over millions of years.

6. Earth's lithosphere is made of huge pieces of solid rock called ____________________ .

7. These solid pieces of rock rest on the hot, soft, slippery rock of Earth's ____________________ .

8. Melted rock called ____________________ rises up through the crack where plates move apart under the ocean.

9. As the ocean floor spreads at the plate boundary, the ____________________ resting on the plates also move apart.

Name ______________________ Date ____________

10. As the ridge grows higher, a(n) ______________ valley forms along the center of it.

How does the movement of Earth's plates affect the land?

11. When plates push toward each other, a force called ______________ results.

12. Because of this force, the ground at the edges of plates is pushed upward to form ______________ mountains.

13. A mountain range in Asia, the ______________, began to form in this way millions of years ago.

14. In places where one plate rubs past another, a twisting or tearing force called ______________ results.

15. This force can cause blocks of crust to break apart along deep cracks in Earth's crust called ______________.

16. When rock on one side of a fault moves down and rock on the other side moves up, a ______________ mountain is formed.

17. A California mountain range, the ______________, is this type of landform.

Critical Thinking

18. Compare how two types of mountains are formed.

____

____

____

____

Name ______________________________ Date ____________

Plate Tectonics

Use the terms in the box below to fill in the blanks.

continental drift	mid-ocean ridge
fault-block mountains	plate tectonics
folded mountains	rift valley
geologist	

1. As hardened magma builds up on both sides of a plate boundary, a(n) ______________________ forms at the center of a mid-ocean ridge.

2. The force of compression can form ______________________ at the point where two plates push together.

3. The theory that states that the continents were once one landmass and that they drifted to their present positions over many years is called ______________________ .

4. A highland in the middle of the oceans that runs parallel to the continents is called a(n) ______________________ .

5. A scientist who studies Earth's structure and history is called a(n) ______________________ .

6. Shear forces at a fault can form ______________________ .

7. The model that states that Earth's surface is composed of large rock plates that fit together like jigsaw puzzle pieces is called ______________________ .

LESSON
Cloze Activity

Name ______________________ Date ____________

Plate Tectonics

Fill in the blanks.

compression	continents	fossils	shear
continental drift	folded	plate tectonics	

The continents were not always where they are today. About 100 years ago, Alfred Wegener developed the theory of ________________. The theory states that Earth's ________________ were once one landmass. The landmass broke up millions of years ago, and the continents drifted to the positions we know today. Wegener supported his theory with evidence from rocks and ________________. Later, scientists developed the theory of ________________.

When plates push together, they produce the force of ________________. This force can push the ground at the boundary upward, forming ________________ mountains. When plates slide past each other, they create ________________. This force can make huge blocks of crust break apart along faults. Over millions of years, the blocks can shift upward to form fault-block mountains.

Name ______________________ Date ____________

Writing in Science

Pangea and Other Supercontinents

Write About It

What evidence do scientists have that Rodinia and Pannotia existed? Research this evidence and select a main idea. Write an expository essay with details to support your main idea.

Getting Ideas

Do some research to find out whether Rodinia and Pannotia actually existed. Use the chart below. In the boxes on the top, write details that you find. In the box on the bottom, summarize this information.

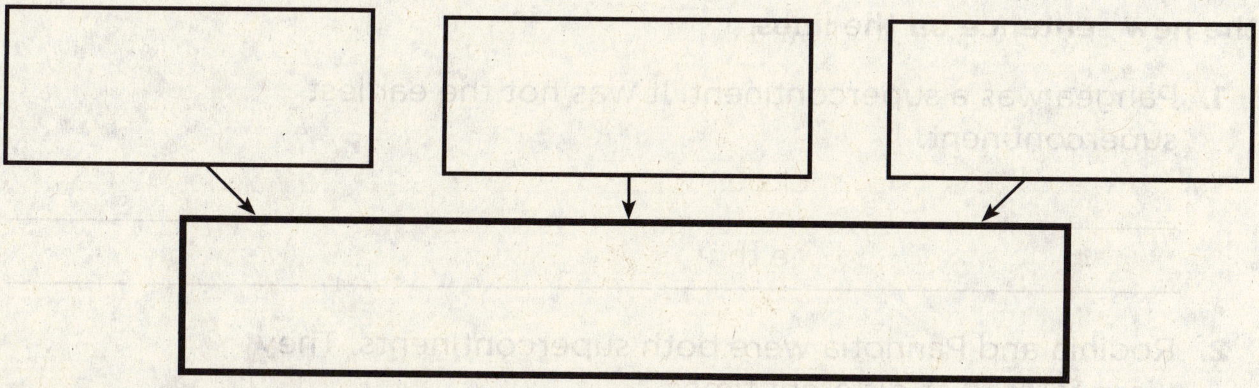

Planning and Organizing

Here are two sentences that Mai wrote. Write Yes if the sentence supports the idea that Rodinia and Pannotia actually existed. Write No if it does not.

1. There are common rock types and structural features along the coastlines of continents today. ________
2. Figuring out how supercontinents formed and broke apart is a lot like detective work. ________

Name ______________________ Date ____________

Drafting

Write a sentence to begin your essay. This sentence should tell your main idea about Rodinia and Pannotia.

____

____

____

Review the evidence you found and your summary. Now write the first draft of your essay. Use a separate piece of paper. Include facts and details that back up your main idea. Draw a conclusion at the end.

Revising and Proofreading

Help Mai revise her writing. Use the word but to combine each pair of sentences. Put a comma before this word. Write the new sentence on the lines.

1. Pangea was a supercontinent. It was not the earliest supercontinent.

____

____

2. Rodinia and Pannotia were both supercontinents. They were formed at different times.

____

____

Now revise and proofread your writing. Ask yourself:

- Did I clearly state my main idea?
- Did I include facts and details to back up my idea?
- Did I reach a sound conclusion at the end?
- Did I correct all mistakes?

Name ______________________ Date __________

LESSON
Outline

Volcanoes

Use your textbook to help you fill in the blanks.

Where are volcanoes found?

1. Most of Earth's volcanoes are located at edges of ______________________ .
2. A string of volcanoes at plate boundaries around the Pacific Ocean is known as the ______________________ .
3. Volcanoes often erupt at places where one plate ______________________ the other.
4. The bottom edge of the diving plate melts in the heat of the ______________________ .
5. The melted rock rises within the crust, forming a hot pool of ______________________ .
6. The hot rock sometimes erupts through openings in Earth's surface as a(n) ______________________ .

How do volcanoes build land?

7. When magma hardens inside Earth's crust, it can form vertical ______________________ and horizontal sills.
8. Magma pushed into a thick sill can form a(n) ______________________ .
9. The largest underground magma formations are ______________________ , which can form large hills.
10. Magma that reaches Earth's surface is ______________________ .

LESSON **Outline**

Name ______________________________ Date ______________

11. A volcano that is ________________ can erupt with lava, ash, gas, or rock.

12. When a volcano stays quiet for a time, it is ________________.

13. A volcano that no longer erupts is ________________, or dead.

How do volcanoes build islands?

14. The Hawaiian Islands formed over a stationary pool of magma below Earth's crust called a(n) ________________.

15. When the mountains grew high enough to break the ocean's surface, they became volcanic ________________.

16. As the plate moved slowly from southeast to northwest, the Hawaiian Island ________________ formed.

17. Where two ocean plates meet and one is pushed under the other, an island ________________ may form.

18. Magma from edge of the lower plate rises and builds volcanic islands along the plate ________________.

19. An example of an island arc is the ________________ in Alaska.

Critical Thinking

20. Why do volcanoes form when one plate dives under another?

____

____

____

Name ______________________ Date ____________

Volcanoes

Match the correct letter with the description.

a. cinder-cone volcano	**f.** island chain
b. composite volcano	**g.** lava
c. crater	**h.** shield volcano
d. hot spot	**i.** volcano
e. island arc	

1. ________ magma that reaches Earth's surface

2. ________ a series of volcanic islands that form along a plate boundary

3. ________ a broad volcano with gently sloping sides formed from thin, fluid lava

4. ________ an opening in Earth's crust through which magma flows

5. ________ a stationary pool of magma below Earth's crust

6. ________ a large, cone-shaped volcano built from alternating layers of cinders and hardened lava

7. ________ a line of islands

8. ________ a cup-shaped depression that forms around a volcano's vent

9. ________ a cone-shaped volcano of cinders, with a narrow base and steep sides

Name ______________________ Date ____________

Volcanoes

Fill in the blanks.

cinder-cone	lava	plates	volcano
composite	mantle	shield	

Openings on the Earth's surface appear on the edges of the crust's plates. An opening in Earth's crust from which magma flows is a(n) ________________ . Most volcanoes form in places where ________________ push toward each other, and one dives under the other. The lower edge of the diving plate melts in the ________________ , producing hot magma that rises in the crust. Magma that breaks through to Earth's surface is ________________ .

There are three types of volcanic mountains. A large, broad mountain composed of hardened lava is a(n) ________________ volcano. A narrow, steep mountain formed from cinders is a(n) ________________ volcano. A large, cone-shaped mountain formed by layers of lava and cinders is a(n) ________________ volcano. Volcanoes are built up over time as more material is deposited.

Name ______________________ Date __________

LESSON
Outline

Earthquakes

Use your textbook to help you fill in the blanks.

What is an earthquake?

1. Earthquakes happen when the layers of rock on both sides of a(n) ______________ suddenly slip.

2. Waves of energy spread out from the ______________, the place where the slipping began.

3. When they reach the surface, waves spread out from the ______________ of the earthquake (the point directly above the focus).

4. Most earthquakes happen at faults that are near the boundaries of ______________.

What waves are produced in an earthquake?

5. Scientists use a(n) ______________ to detect and measure earthquake waves.

6. The fastest earthquake waves, ______________ waves, pass through solids and liquids and move back and forth.

7. An earthquake's ______________ waves do the most damage because they move up and down and from side to side.

8. The slowest-moving waves, ______________ waves, move across Earth's surface like ripples on a pond.

How are earthquakes measured?

9. Scientists use the ______________ scale to measure earthquake magnitude.

10. A measure of the amount of ______________ that an earthquake releases is magnitude.

Name ______________________ Date __________

11. Scientists use the ______________ scale to measure an earthquake's effects.

12. An underwater earthquake can produce a large wave called a(n) ______________ .

13. Underwater earthquakes with a magnitude of ______________ or greater on the Richter scale are most likely to cause tsunamis.

How can people prepare?

14. Layers of rubber and steel between a building and its foundation allow the building to ______________ , reducing the damage caused by up-and-down motions.

15. Before an earthquake, people should secure heavy or tall furniture to ______________ to prevent it from falling and causing injury.

16. In their attempt to tell when earthquakes might happen, scientists look for possible warning signs such as changes in the angle of the ______________ .

17. Earthquakes are hard to ______________ , but the ability to do so would allow early warnings that could save lives.

Critical Thinking

18. Which scale do you think would better explain an earthquake to you, the Richter scale or the Mercalli scale? Why?

____

____

____

Name ______________________ Date __________

Earthquakes

Use the clues below to find the words hidden in the puzzle.

S	M	G	A	R	J	F	D	H	O
Z	M	F	O	I	G	A	D	B	Y
E	A	R	T	H	Q	U	A	K	E
C	G	C	D	E	L	L	X	R	T
V	N	E	R	G	K	T	I	W	S
R	I	C	H	T	E	R	B	O	P
G	T	S	U	N	A	M	I	Y	U
V	U	H	I	P	F	O	C	U	S
X	D	G	B	E	C	K	L	T	T
Z	E	P	I	C	E	N	T	E	R

1. A sudden movement of Earth's crust is a(n) ______________________ .
2. The point on the surface directly above an earthquake's focus is its ______________________ .
3. A crack in Earth's crust is a(n) ______________________ .
4. The place along a fault where the slipping that causes an earthquake begins is the earthquake's ______________________ .
5. A measure of the energy that an earthquake releases is its ______________________ .
6. A large ocean wave caused by an underwater earthquake is a(n) ______________________ .
7. The scale that measures the magnitude of an earthquake is called the ______________________ scale.

Name ______________________ Date ____________

Earthquakes

Fill in the blanks.

earthquake	primary or P
energy	Richter
fault	secondary or S
Mercalli	

The plates of the Earth are in motion. A sudden movement of Earth's crust is a(n) ______________ . Most earthquakes occur near plate boundaries, when layers of rock that usually adhere to each other suddenly slip at a(n) ______________ . The scale that measures the magnitude of an earthquake is called the ______________ scale. The scale that measures how severe an earthquake feels and the amount of damage the quake does to objects is called the ______________ scale.

The movement of plates during an earthquake sends out waves of ______________ that shake the ground. When an earthquake occurs, ______________ waves move back and forth very rapidly. An earthquake's ______________ waves do the most damage because they move both up and down and from side to side. The slowest waves are surface or L waves.

Name ______________________________ Date ______________

How Earthquakes Help Predict Volcanic Eruptions

Write About It

What are the differences between short-period and long-period earthquakes? Research these earthquakes. Write an explanatory essay with details that support your main idea.

Getting Ideas

Find out more about these types of earthquakes. Use the chart below to record information. Write the cause of each type of earthquake in the left-hand box. Write the type of earthquake in the right-hand box.

Cause	→	Effect
	→	
	→	
	→	

Planning and Organizing

Hector wants to write first about short-term earthquakes, then about long-term earthquakes, and finally about harmonic-tremor earthquakes. Here are three sentences he wrote. Help him organize them. Then write 1 by the sentence that should come first. Write 2 by the sentence that should come second. Write 3 by the sentence that should come last.

 Name ______________________ Date __________

1. ________ Seismographs can detect the ongoing flow of magma in a harmonic-tremor earthquake.

2. ________ A short-term earthquake may be too small to be felt.

3. ________ When the gas builds up, you have a long-term earthquake.

Drafting

Write a sentence to begin your explanation. Tell your main idea about the types of earthquakes you have researched.

____

____

Now write your explanatory essay on a separate piece of paper. Tell how each type of earthquake occurs. Use time-order or spatial words to make your explanation clear.

Revising and Proofreading

Here is part of Hector's explanatory essay. Add a time-order word or a spatial-order word in each blank to make the meaning clearer.

________________ magma pushes its way ________________ to the surface, the rocks begin to break. ________________ long-term earthquakes begin. Finally, magma flows ________________ the surface and breaks through.

Now revise and proofread your writing. Ask yourself:

- Did I discuss each type of earthquake and explain the difference?
- Did I include clear details that are easy to follow?
- Did I include time-order words or spatial order words?
- Did I correct all mistakes?

Name ______________________ Date ____________

LESSON
Outline

Shaping Earth's Surface

Use your textbook to help you fill in the blanks.

What is weathering?

1. The process that breaks down rock into small pieces is called ________________ .
2. Impacts, temperature changes, and ice expanding in cracks break down rock in the process of ________________ .
3. When rock's composition is broken down and changed, ________________ occurs.

What is erosion?

4. Pieces of weathered rock are moved from place to place by ________________ .
5. When rock and soil on a slope become loose, gravity can move them downhill in a ________________ .
6. A large mass of flowing ice, called a ________________ , can erode the rock and soil beneath it.
7. The process of ________________ picks up eroded material and leaves it in a different place.

How do erosion and deposition work together?

8. The running water of ________________ erodes rock and soil and washes it downstream.
9. Slow-moving rivers can flow in gentle loops called ________________ .

Name ______________________ Date ____________

10. Rivers slow down and deposit sediment at their mouths in fan-shaped landforms called ________________.

11. Waves can move sand along beaches or deposit it offshore to build strips of sand called ________________.

How are shorelines changed?

12. During floods, rivers deposit sediment on ________________ along their banks.

13. Deposits of sand along the shore can form ________________ islands that protect the beaches behind them from storm waves.

14. Wind deposits sand along the shoreline in hills of sand called ________________.

How can shorelines be protected?

15. People build walls called ________________ along rivers to prevent floods.

16. Natural ________________ along rivers soak up water and reduce the chance of flooding.

17. Fences and ________________ protect dunes by preventing sand from blowing away.

Critical Thinking

18. How can people help stop erosion on a beach?

____

____

____

____

____

Name ______________________________ Date ____________

Shaping Earth's Surface

Who am I? What am I?

Choose a word from the word box below that answers each question.

a. acid rain	**d.** erosion	**g.** meander
b. delta	**e.** floodplain	**h.** sediment
c. deposition	**f.** glacier	**i.** weathering

1. ________ I am a huge mass of ice. I erode the rock beneath me as I flow over it. Who am I?

2. ________ I am particles of rock. Moving water deposits me when it slows down. What am I?

3. ________ I wear away stone and metal surfaces when I fall from the sky. What am I?

4. ________ I am a flat area along a riverbank. Rivers deposit sediment on me when they flood. Who am I?

5. ________ I drop eroded sediment in a different place after I picked it up. What am I?

6. ________ I am a fan-shaped piece of land. I form from deposits at the mouths of rivers. Who am I?

7. ________ I break down rock into smaller pieces. What am I?

8. ________ I am a gentle loop in a slow-flowing river. What am I?

9. ________ I pick up and move pieces of weathered rock. What am I?

Name ____________________ Date ____________

Shaping Earth's Surface

Fill in the blanks.

beaches	erosion	physical weathering
chemical weathering	glaciers	waves
deposition	landslides	weathering

Rocks are constantly being changed by the environment. Over long periods of time, the process of ________________ breaks down rock. When ________________ occurs, the rock simply breaks into smaller pieces. During ________________, the rock weakens as the minerals in it are changed. After weathering, ________________ removes the weathered rock. Gravity pulls loosened rock downhill in ________________. Erosion and deposition shape shorelines. Ocean ________________ and currents move sediment along shorelines. They erode sand from some ________________ and deposit it on others.

As ________________ move over the ground, they scoop out depressions and move the loose rock beneath them. Water, wind, and ice can drop eroded materials in a different place in a process called ________________. This process also changes landforms.

Name ______________________ Date ____________

Wrestling with the Big Muddy

Read the Reading in Science feature in your textbook.

Problem and Solution

Identify the main problem described in the reading. Then identify the solution to the problem and the steps taken to reach it. Write the information in the correct boxes in the graphic organizer below.

Problem

↓

Step to Solution

↓

Step to Solution

↓

Solution

Reading in Science

Name ______________________ Date ____________

Planning and Organizing

Read the "Write About It" questions carefully. Find the text within "Wrestling with the Big Muddy" that answers each question. Use the questions below to help organize your information.

List structures that now control the Missouri River.

a. ______________________

b. ______________________

List the problems the new structures cause.

a. ______________________

b. ______________________

Drafting Your Answers

Now, use the information above to help you write answers to the questions.

1. What did the government do to control the flow of the Missouri River?

____

____

____

2. What problems were caused by building dams along the river?

____

____

____

Name ____________________ Date __________

Our Dynamic Earth

Choose the letter of the best answer.

1. Which of these is a physical feature on Earth's surface?

a. tsunami

b. landform

c. mantle

d. hotspot

2. The crust and the top part of the mantle make up the

a. atmosphere.

b. hydrosphere.

c. asthenosphere.

d. lithosphere.

3. Earth's surface layer is the

a. mantle.

b. crust.

c. biosphere.

d. asthenosphere.

4. What layer of Earth's interior lies just below the crust?

a. mantle

b. inner core

c. lithosphere

d. outer core

5. The plate tectonics model states that Earth's crust is composed of

a. one solid piece of rock.

b. both liquid and frozen water.

c. several huge slabs of rock that fit together.

d. hot, melted rock.

6. What is a fault?

a. energy that an earthquake produces

b. the opening in a volcano

c. a large crack in Earth's crust

d. the boundary between two plates

7. Huge slabs of rock moving suddenly against each other in the Earth's crust create

a. earthquakes.

b. abyssal plains.

c. volcanos.

d. a trench stretch.

Name ______________________ Date ____________

8. A volcano is

   a. an opening in Earth's crust through which magma flows.

   b. any mountain near a plate boundary.

   c. a group of faults near a hot spot.

   d. movement at a fault.

9. The low area between mountains is called a

   a. plateau.

   b. trench.

   c. valley.

   d. landform.

10. The wide, flat are of the ocean floor is known as the

    a. mantle.

    b. abyssal plain.

    c. trench stretch.

    d. aquatic plateau

11. What is the term used for melted rock that reaches the Earth's surface?

    a. lava

    b. mantle

    c. magma

    d. boundary rock

12. This device is used to detect and measure Earthquake waves.

    a. wavometer

    b. richtometer

    c. barometer

    d. seismometer

13. This mass of large flowing ice can erode rock.

    a. glacier

    b. landslide

    c. iceberg

    d. delta

14. Underwater earthquakes of a great magnitude can create

    a. continental divides.

    b. trenches.

    c. tsunamis.

    d. aquatic drift.

Name ______________________ Date __________

CHAPTER
Concept Map

Georgia's Features

Complete the concept map with information you have learned about the regions of Georgia. Some parts have been done for you.

Region	How the Region Formed	Type of Rocks and Sediments	Water Features	Any Unique Features
Appalachian Plateau	______ ______	______ ______	______ ______ ______	______ ______
Valley and Ridge	______ ______ ______	______ ______	steep and fast-flowing rivers; waterfalls	______ ______
Blue Ridge	______ ______ ______	______	______ ______ ______	______ ______
Piedmont	______ ______ ______	______	rivers, reservoirs, small lakes	Stone Mountain
Coastal Plains	deposition of sediments	sedimentary	______ ______	______ ______

LESSON **Outline**

Name ______________________ Date __________

Exploring Georgia's Landforms

Use your textbook to help you fill in the blanks.

What are the western Appalachian Mountains?

1. If you were to tour the state beginning in the far northwest you would see flat-topped mountains of the ______________ region.
2. The ______________ marks the gradual change from flat lands in the north to the Valley and Ridge region in the west.
3. Formations found in the ______________ region are made up mostly of sedimentary rock.
4. Ancient ______________ produced ash that can be found today as a type of clay in Georgia.
5. A total of three tectonic collisions involving North America occurred, with the last collision forming the supercontinent known as ______________.
6. The ______________ was formed when a collision pushed rocks from the Piedmont on top of the rock found in the Valley and Ridge.

What are the eastern Appalachian Mountains and the Piedmont?

7. Many of the rocks found in the Blue Ridge are ______________, which is rock that changes in form.
8. The first United States gold rush began in 1828 at ______________.
9. The highest peak in Georgia, ______________, is made up of soapstone and other minerals.

Name ______________________________ Date ____________

10. The rolling hills of the ______________________ region are the remains of heavily eroded mountain peaks.

11. The highest point in the Appalachians today is ______________________ in North Carolina at 6,684 ft.

What is the Coastal Plain?

12. The sedimentary rock layers of the ______________________ region contain fossils from ancient marine organisms dating back to 145–65 million years ago.

13. The remains of many large organisms have been found in sedimentary rock, including giant crocodiles, duck-bill dinosaurs, and even a relative of the *Tyrannosaurus rex* named ______________________.

14. Numerous animal shells can form a type of sedimentary rock with many small holes called ______________________.

15. The ______________________, or layer of water-containing rock, provides water to the Coastal Plain.

16. Where water flows above the ______________________ you will find waterfalls and rapids.

Critical Thinking

17. Why are the Appalachian Mountains currently much shorter and rounder than the Himalayas?

____

____

____

____

Name ______________________ Date ____________

Exploring Georgia's Landforms

Match the correct letter with the description.

a. Appalachian Plateau region	**e.** Blue Ridge region
b. Cumberland Plateau	**f.** Piedmont region
c. Valley and Ridge region	**g.** Brevard Fault Line
d. Cartersville Fault	**h.** Coastal Plains region

1. ________ a feature separating the Piedmont region from the Blue Ridge region.

2. ________ an area of Georgia just east of the Valley and Ridge region and made of many of the same types of rock.

3. ________ a feature separating the Piedmont region from the Valley and Ridge region.

4. ________ an area in the far northwest of Georgia with flat-topped mountains.

5. ________ the low-lying eastern portion of Georgia with layers of sedimentary rock.

6. ________ an area of Georgia characterized by its rolling hills that were once mountain peaks.

7. ________ an area of Georgia containing sandstone that has been heavily eroded, resulting in such landforms as Sand Mountain and Lookout Mountain.

8. ________ an area of Georgia characterized by soaring peaks and ridges separated by long wide valleys.

Name ______________________ Date __________

LESSON
Cloze Activity

Exploring Georgia's Landforms

Fill in the blanks.

Appalachian Plateau	Cumberland Plateau
Blue Ridge	sedimentary rock
Cartersville Fault	supercontinents

Georgia has many different landforms. In the northwestern part of the state, the ____________________ region is home to many flat-topped mountain peaks. The heavily eroded sandstone of the ____________________ marks the transition from the mountains into the peaks and wide valleys of the Valley and Ridge region. The valleys and ridges formed as different types of ________________ eroded at different rates. The folds in the region were created as ________________ collided over millions of years.

Other collisions led to the formation of the ________________, which separates the Valley and Ridge region from the Piedmont region. The high pressure of collisions also helped create the metamorphic and igneous rock of the ________________ region. Continent movement also caused the formation and destruction of ancient seas, which placed many fossils in the sedimentary rock of the Coastal Plains region.

LESSON **Outline**

Name ______________________ Date ____________

Exploring Georgia's Water Features

Use your textbook to help you fill in the blanks.

What are Georgia's rivers and waterfalls?

1. The ______________ of an area determines how the water will flow through it.

2. All of the land that drains into a river is that river's

______________ .

3. The ______________ is the major river in Georgia flowing to the Atlantic Ocean.

4. The point at which waterfalls stop appearing on a river is the

______________ .

5. The major river flowing to the west in Georgia is the

______________ .

6. Waters flowing from the Amicalola Mountains rush over a ledge in the southern end of the Blue Ridge region to form

______________ .

What are Georgia's lakes and swamps like?

7. A natural lake formed by the erosion of winding rivers is a(n)

______________ .

8. Underground rock that dissolves can form a(n)

______________, which can fill with water over time and form a natural lake.

9. Artificial lakes, or ______________, are used for flood control, hydroelectric power, drinking water, and human recreation.

Name ______________________ Date ____________

LESSON Outline

10. Power companies build dams for cooling and ____________ power.

11. ________________ and Lake Sidney Lanier provide flood protection and drinking water for Atlanta.

12. Before you can drink water from a reservoir, it must be __________.

What is a swamp?

13. The largest swamp in North America is the ____________________.

14. The spongy ground helps with ______________ control.

15. Swamps have nutrient-poor and acidic soil, which allow

for the growth of ________________ because they are adapted to grow there.

16. Swamps like ________________ near Augusta help filter the city's water supply.

Critical Thinking

17. What are three ways that water features in Georgia are beneficial?

____

____

____

____

Name ______________________ Date ____________

Exploring Georgia's Water Features

Match the correct word with its description, and fill in the crossword puzzle.

dam	oxbow lake	swamp
fall line	reservoir	watershed

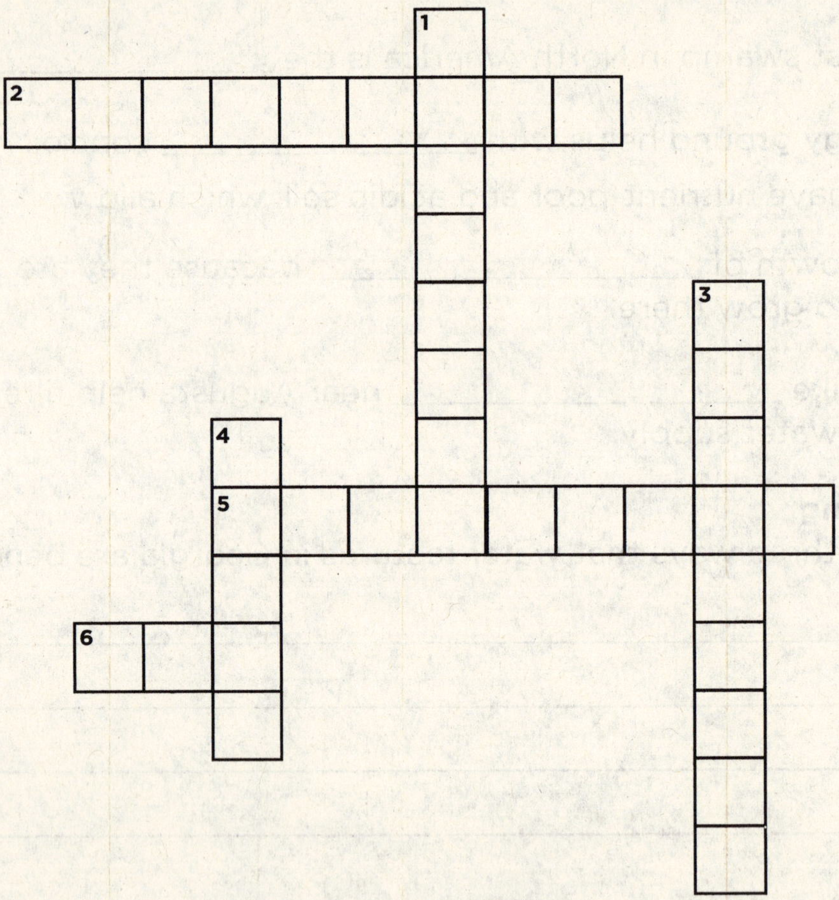

Across

2. lake formed from a portion of a stream channel that is cut off from the rest of the stream by erosion
5. area of land that contributes to a river or river system.
6. structure built to regulate the flow of water

Down

1. the point at which waterfalls stop appearing on rivers
2. artificial lake created by building dams that are used for hydroelectric power generation and flood control
3. wetland area with shrubs and trees and wet, spongy ground

Name ____________________ Date ____________

LESSON
Cloze Activity

Exploring Georgia's Water Features

Fill in the blanks.

adaptations	hydroelectric power	waterfall
Atlantic Ocean	reservoirs	west
ecosystem	swamp	

There are two main groups of rivers in Georgia. One group flows east toward the ________________. The other group flows ________________ toward the Gulf of Mexico. There are many plateaus and mountains in the northern part of the state. There are many ________________ in this part of the state because of the changes in elevation.

Georgia does not have many large natural lakes. However, there are many large ________________ in the state. These are created by building dams to stop the flow of water. The dams can then be used to generate ________________ for the people who live near the dams.

A ________________ is a wet, spongy area with acidic soil. This type of water feature is an important ________________ that supports a diversity of plants and animals. These organisms have special ________________ to live in this environment.

Name ____________________ Date ____________

Keep Our Water Clean

Write About It

Write a letter to the mayor of your town. Explain a need that the students in your community have and why people should help. State your opinion clearly and support it with relevant facts and evidence organized in a logical way.

Getting Ideas

Think of an issue that clearly affects life in your community. Form an opinion about it. Write this opinion in the top box in the chart below. Then jot down reasons that support this opinion in the bottom boxes.

Opinion		

Reason	Reason	Reason

Planning and Organizing

Kristin's opinion is that the town should ban power boats from the lake. Here are three statements she wrote. Write Yes if the statement supports her opinion. Write No if it does not.

1. ________ Power boats can pollute the water.
2. ________ Power boats are a fun way to spend time on the lake.
3. ________ Power boats kill fish and other water life.

Name ______________________ Date __________

Drafting

Write a statement to begin your letter. Make sure it clearly tells the issue you are concerned about and states your opinion.

____

____

____

Now write your letter to the mayor on a separate piece of paper. Use the form of a business letter. Start the body of your letter with the sentence you wrote above. Include reasons that will persuade the mayor to support your opinion. End by stating what you think should be done. Remember to be polite and respectful.

Revising and Proofreading

Now revise and proofread your writing. Ask yourself:

- ▶ Did I clearly state my opinion about a need that the students in my community have?
- ▶ Did I provide relevant reasons to support this opinion?
- ▶ Did I correct all mistakes in grammar, spelling, capitalization, and punctuation?

Name ______________________ Date __________

Georgia Features

Choose the letter of the best answer.

1. Which was formed by tectonic plates building mountains?

a. Coastal Plains region

b. oxbow lake

c. Appalachian Plateau region

d. fall line

2. Which is an area from which water drains into a river?

a. watershed

b. fall line

c. oxbow lake

d. reservoir

3. Most of the rocks in the Blue Ridge region are

a. sedimentary.

b. igneous.

c. igneous and sedimentary.

d. metamorphic and igneous.

4. What two regions are divided by the Brevard Fault line?

a. Blue Ridge and Appalachian Plateau

b. Piedmont and Blue Ridge

c. Appalachian Plateau and Coastal Plains

d. Coastal Plains and Blue Ridge

5. The Cartersville Fault was formed by the collision between which two continental plates?

a. North America and Antarctica

b. Africa and North America

c. South America and North America

d. Asia and North America

6. There are many fossil remains of prehistoric plants and animals found in

a. the Coastal Plains region.

b. the dinosaur region.

c. the Piedmont region.

d. the Valley and Ridge region.

Name ______________________ Date ____________

Choose the letter of the best answer.

7. The heavily eroded Cumberland Plateau is home to
   - **a.** Stone Mountain and Sapelo Island.
   - **b.** Lookout Mountain and Sand Mountain.
   - **c.** the fall line and Stone Mountain.
   - **d.** Sapelo Island and the fall line.

8. The fall line is the point where
   - **a.** reservoirs are found.
   - **b.** oxbow lakes are found.
   - **c.** waterfalls are found.
   - **d.** marshes and mudflats are found.

9. An oxbow lake is formed by
   - **a.** man-made dams.
   - **b.** rainwater filling in sinkholes.
   - **c.** erosion and deposition on meandering rivers.
   - **d.** building reservoirs.

10. In what geologic region is Stone Mountain found?
    - **a.** Valley and Ridge region
    - **b.** Blue Ridge region
    - **c.** Coastal Plains region
    - **d.** Piedmont region

11. Which of these is **NOT** a use of reservoirs?
    - **a.** Storing garbage and recycling materials
    - **b.** Generating hydroelectric power
    - **c.** Storing fresh water
    - **d.** Flood control

12. Most of the land formations of the Valley and Ridge region are made of
    - **a.** marble, granite, and limestone.
    - **b.** sandstone, limestone, and shale.
    - **c.** soapstone, marble, and sandstone.
    - **d.** shale, marble, and granite.

Name ______________________ Date ____________

The Many Sides of Diamonds

Read the Literature feature in your textbook.

Write About It

Response to Literature This article describes the formation and use of diamonds. Research additional information about the history of industrial diamonds, how they are formed, and how they are used. Write a report about industrial diamonds. Include facts and details from this article and from your research.

Name ______________________ Date ____________

Comparing Kinds of Matter

Complete the concept map by filling in answers where blanks appear.

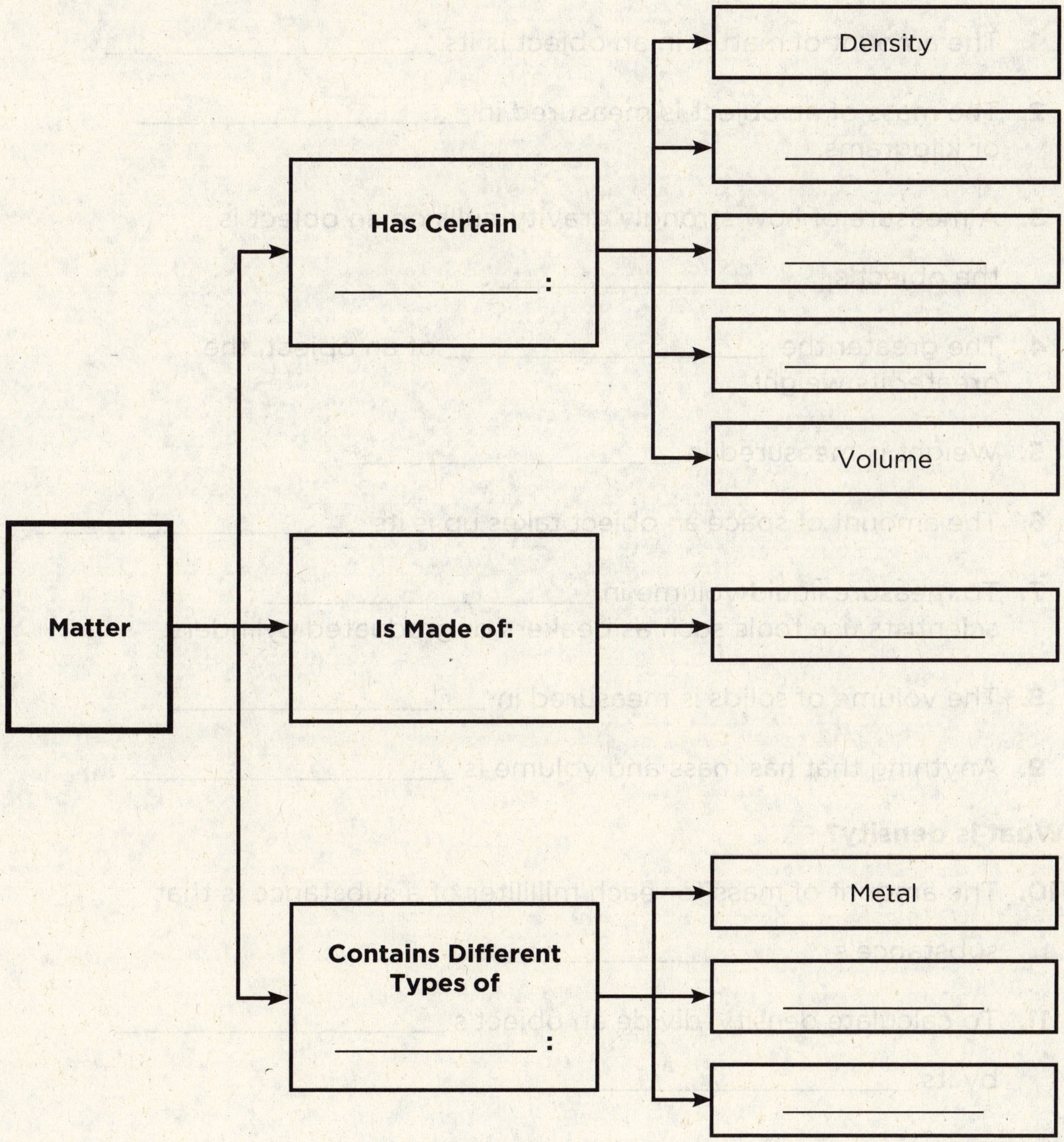

LESSON **Outline**

Name ______________________ Date ____________

Properties of Matter

Use your textbook to help you fill in the blanks.

How can you describe matter?

1. The amount of matter in an object is its ____________.
2. The mass of an object is measured in ____________ or kilograms.
3. A measure of how strongly gravity pulls on an object is the object's ____________.
4. The greater the ____________ of an object, the greater its weight.
5. Weight is measured in ____________.
6. The amount of space an object takes up is its ____________.
7. To measure liquid volume in ____________, scientists use tools such as beakers or graduated cylinders.
8. The volume of solids is measured in ____________.
9. Anything that has mass and volume is ____________.

What is density?

10. The amount of mass for each milliliter of a substance is that substance's ____________.
11. To calculate density, divide an object's ____________ by its ____________.

12. Buoyancy depends on ____________________, which depends on mass and volume.

13. Changing the mass or volume of an object changes its density and ____________________.

14. If an object covers a large enough area of the water's surface, it can float on the water because of the ____________________ of water particles.

What forms can matter have?

15. Matter can exist as a solid, a(n) ____________________, or a gas.

16. A solid has a definite ____________________ and volume.

17. A liquid has a definite ____________________, but it takes the shape of the container holding it.

18. A gas does not have a definite volume or a definite ____________________.

Critical Thinking

19. How can matter be described?

____

____

____

____

____

____

____

____

Name ______________________ Date ____________

Properties of Matter

Fill in the crossword puzzle from the clues below.

buoyancy	mass	Newton	volume
density	matter	surface tension	weight

Across

4. The amount of space that matter takes up
5. The metric unit used to measure weight
6. Anything that has mass and volume
7. The amount of mass for each milliliter of a substance

Down

1. The property of water that helps certain objects float
2. An object's resistance to sinking
3. How strongly gravity pulls on an object
6. The amount of matter in an object

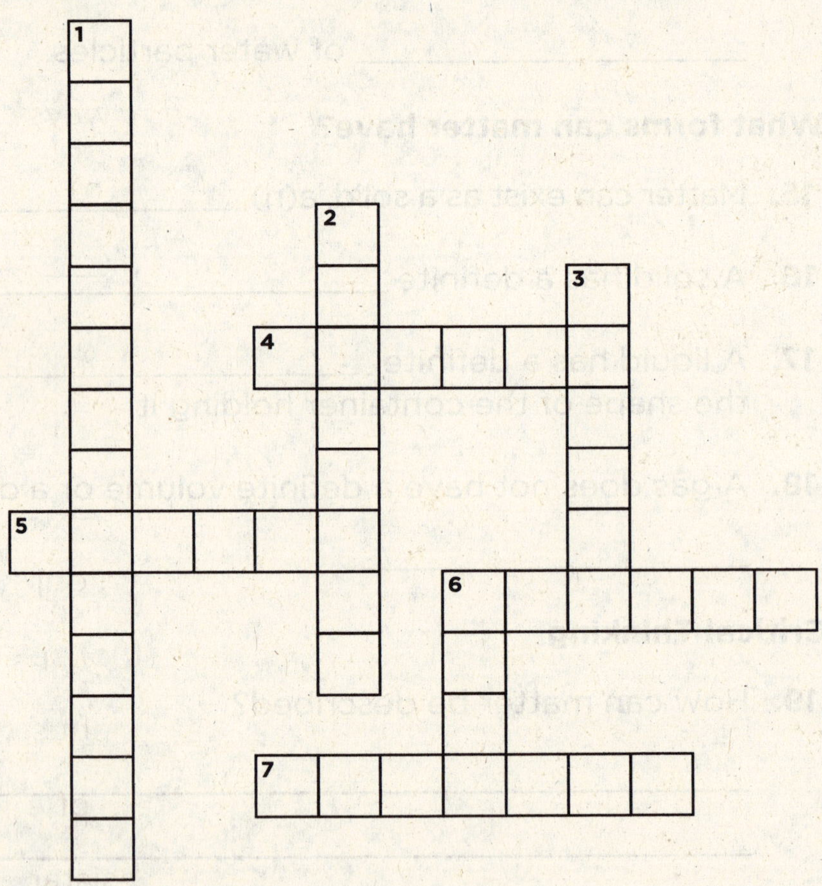

Name ______________________ Date ____________

Properties of Matter

Fill in the blanks.

buoyancy	float	properties	solid
constant	gas	push	volume
density	mass	sink	weight

We describe matter in a number of ways. Matter can exist as a(n) ______________, a liquid, or a(n) ______________. Scientists use these and other ______________ to identify matter.

The amount of matter in an object is the object's ______________, a property that is ______________. However, the ______________ of an object changes as the force of gravity changes. The amount of space that an object takes up is its ______________. Scientists also measure the amount of matter for each milliliter of a substance, or its ______________. An object's resistance to sinking is ______________. When an object is placed on a fluid, the object and the fluid ______________ against each other. If the fluid is denser, the object will ______________. If the object is denser, the object will ______________. Matter is anything that has mass and volume.

Name ______________________________ Date ____________

Elements

Use your textbook to help you fill in the blanks.

What is matter made of?

1. A substance that cannot be broken down chemically into simpler substances is a(n) ______________________ .
2. One important property of an element is its ______________________ at room temperature.
3. Another important property of an element is the way that it ______________________ with other elements.
4. Today we know that a(n) ______________________ is the smallest unit of an element that has that element's properties.

What are atoms and molecules made of?

5. The center of an atom is its ______________________ .
6. An atom's nucleus contains particles called protons that have a positive charge and particles called ______________________ that have no charge.
7. Negatively charged particles called ______________________ move around the nucleus.
8. Because an atom has the same number of ______________________ and electrons, the atom has no overall charge.
9. The number of protons in an atom is that atom's ______________________ .

Name ______________________ Date ____________

10. An atom's protons and neutrons have about the same mass, which is one ______________ unit, or a.m.u.

11. The mass of all particles of an atom added together is its ______________ .

12. Two or more atoms joined into a single particle form a(n) ______________ .

13. Molecules have properties that are different than the ______________ that form them.

How are elements grouped?

14. Dmitri Mendeleev created the ______________ of elements.

15. The table's columns group elements according to their ______________ .

What are the most common elements?

16. In space, the most common elements are ______________ and helium.

17. On Earth, elements such as ______________ and any of these: oxygen, silicon, aluminum, nitrogen, iron, and calcium are among the most common.

Critical Thinking

18. What is matter made of?

____

____

____

LESSON
Vocabulary

Name ______________________ Date ____________

Elements

Read each clue. Write the answer in the blanks using the words below.

atom	element	molecule	nucleus
electron	metal	neutron	proton

1. The smallest unit of an element that retains that element's properties is a(n) ____________________ .

2. The particle in an atom that has a negative charge is a(n) ____________________ .

3. A substance that chemical reactions cannot break down into something simpler is a(n) ____________________ .

4. An element that has properties such as shine, conductivity, and flexibility is a(n) ____________________ .

5. Two or more atoms that are joined into one particle are a(n) ____________________ .

6. In the nucleus of an atom, a particle that has no electrical charge is a(n) ____________________ .

7. The center of an atom is its ____________________ .

8. In the nucleus of an atom, a particle that has a positive electrical charge is a(n) ____________________ .

Name ______________________ Date ____________

LESSON
Cloze Activity

Elements

Fill in the blanks.

atoms	metal	nonmetal	properties
electrons	metalloid	nucleus	temperature
elements	neutrons	periodic table	

Every substance on Earth is made of one or more ______________. Dmitri Mendeleev created the ______________ in the 1860s. It groups elements according to their ______________. One important property of an element is its state at room ______________. Another is the way that it combines or mixes chemically with other elements. A third property is the element's classification as a(n) ______________, a(n) ______________, or a(n) ______________.

Each element is composed of tiny particles called ______________, the smallest units that retain the element's properties. All atoms have the same parts. The center of an atom is its ______________. The nucleus contains protons and ______________. Atoms also contain ______________, which move around the nucleus. Protons and neutrons have a much larger mass than electrons do.

Name ______________________ Date ____________

Element Discovery

When Mendeleev shuffled his element cards to create the periodic table in 1869, he suspected he wasn't playing with a full deck. Many of the elements had already been discovered, but he believed others would come later.

1766 Hydrogen—The most abundant atom in nature is discovered by Henry Cavendish. In 1766, Cavendish is experimenting with materials in his lab when he isolates a gas that is flammable. He realizes that this gas might be a new element and calls it flammable air. The element later gets its name from the Greek words meaning "water forming," when another scientist discovers that water is made of hydrogen and oxygen.

1772–74 Oxygen—Scientists Joseph Priestley and Carl Wilhelm Scheele independently discover that when they heat certain compounds, a new kind of "air" or gas is given off. The new gas makes substances burn five times faster than ordinary air. The new gas is named oxygen from the Greek words meaning "acid former." That's because when oxygen combines with other elements, the compounds are usually acidic.

1868–1895 Helium—Joseph Lockyer discovers helium in 1868 by studying the Sun's spectrum with a spectroscope during a solar eclipse. He finds color lines that no element at the time was known to produce. He infers the lines must be due to a new element found only in the Sun. The element is named helium, after Helios, the Greek god of the Sun. In 1895, helium is finally found on Earth in uranium minerals.

1940 Plutonium—Scientists in Berkeley, California, create a new element by bombarding uranium with particles of deuterium, a special form of hydrogen. They name the element after the recently discovered planetary body Pluto.

Name ______________________ Date ____________

1952 Einsteinium—A team of scientists find this element while studying the radioactive debris created when a hydrogen bomb explodes. They name it in honor of scientist Albert Einstein. Only a small amount of einsteinium has ever been produced, and it exists only for a fraction of a second before it transforms itself into other elements.

The periodic table isn't finished. Elements are still being added to it. In the past 75 years, 26 new elements have been added to the table. That's about one element every three years! If you found a new element, what would you name it?

Write About It

Classify

1. Which elements were discovered as gases?
2. Which elements have names that describe their properties? How are the other elements named?

Name ______________________ Date __________

Changes of State

Use your textbook to help you fill in the blanks.

How can matter change state?

1. Altering the form or organization of an object without changing the type of matter within it is called a(n) ____________________.
2. The three states of matter are ____________________, liquid, and ____________________.
3. The state of matter of an object is a(n) ____________________ property.
4. The average vibration of molecules in an object is measured by ____________________.
5. When a solid gains heat energy, its molecules begin vibrating too quickly to stay together, so the solid becomes a(n) ____________________.
6. When gases lose heat, they ____________________ into liquids.
7. A liquid loses heat and ____________________ into a solid.
8. When a solid changes directly into a gas, it ____________________.
9. Most liquids become ____________________ when they change to a solid.

Name ______________________________ Date ____________

LESSON
Outline

When does matter change state?

10. When a substance melts or boils, it absorbs ____________________.

11. The temperature at which a substance changes from a solid to a liquid is its ____________________.

12. The temperature at which a substance changes from a liquid to a gas is its ____________________.

13. The temperature at which a substance changes from a liquid to a solid is its ____________________.

14. Nonmetals are weakly attracted to one another, so they have ____________________ melting and boiling points.

15. The slow change from a liquid to a gas at temperatures below the boiling point is called ____________________.

What are expansion and contraction?

16. An increase in an object's volume when it is heated is called ____________________; a decrease in its volume when it is cooled is called ____________________.

Critical Thinking

17. How does water change when heat is added or removed?

____

____

____

____

____

Name ______________________ Date ____________

Changes of State

Choose words from the word box below to finish the crossword puzzle.

boiling	freezing	sublimation
contraction	melting	

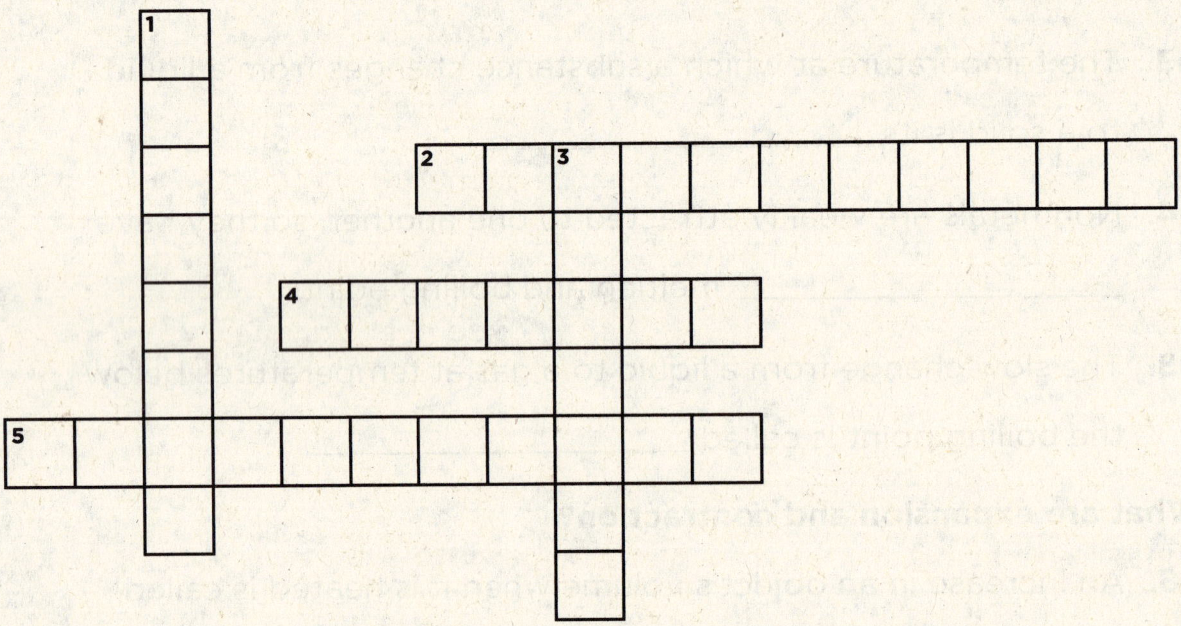

Across

2. A change from a solid to a gas.
4. The temperature at which water changes from a solid to a liquid is the ______________ point.
5. A decrease in an object's volume because of a change in temperature is thermal ______________.

Down

1. The temperature at which water changes from a liquid to a solid is its ______________ point.
3. The temperature at which water changes from a liquid to a gas is the ______________ point.

Name ______________________ Date __________

LESSON
Cloze Activity

Changes of State

Fill in the blanks.

boiling point	heat energy	solid
freezing point	liquid	sublimation
gas	melting point	temperature

All substances have three common forms called physical states. These states are ______________, liquid, and ______________. The physical state of matter is changed when ______________ is added or taken away. A measure of the average heat energy that a substance has (the average vibration of its molecules) is its ______________. When a solid is heated to its ______________, its molecules start moving faster, and the solid changes into a(n) ______________. When the liquid is heated to its ______________, its molecules move even faster, and the liquid turns into a gas. The melting point of water is 0°C, and its boiling point is 100°C. Sometimes a solid changes directly into a gas without passing through the liquid state, a process called ______________. When a liquid is cooled to its ______________, it becomes a solid. When a gas is cooled, it condenses and becomes a liquid.

LESSON **Outline**

Name ______________________ Date ____________

Mixtures

Use your textbook to help you fill in the blanks.

What are mixtures?

1. A physical combination of substances that remain the same is a(n) ________________.
2. Mixtures can be ________________ into their original substances.
3. Mixtures with different parts that can be plainly seen with the naked eye are called ________________ mixtures.
4. Mixtures that look smooth to the naked eye but speckled under a microscope are called ________________.
5. Over time, one or more parts of a suspension can ________________.
6. A heterogeneous mixture with parts that do not settle out is called a(n) ________________.

What are solutions?

7. A mixture that looks the same everywhere, even under a microscope, is called a(n) ________________.
8. The part of a solution that is dissolved is the ________________.
9. The part of a solution that dissolves the other substance is called the ________________.

Name ______________________ Date ____________

10. A solution of two or more solids is a(n) ____________________.

11. Because it can dissolve many things, water is called the ____________________.

How can you take mixtures apart?

12. To separate one part of a mixture from another, you can use a(n) ____________________.

13. When two liquids in a mixture have different boiling points, they can be separated by ____________________.

14. Because liquids travel at different speeds through an absorbent paper, they can be separated by ____________________.

How are mixtures used?

15. Cheese, gelatin, marshmallows, and paint are all examples of useful ____________________.

16. Copper is alloyed with zinc to make ____________________.

Critical Thinking

17. Suppose you were to mix together salt, water, and mud. Identify type of mixture you have made. Describe how you could separate the parts of the mixture from one another.

____

____

____

____

____

____

Name ______________________ Date __________

Mixtures

Who am I? What am I?

Choose a word from the word box below that answers each question.

a. alloy	**c.** distillation	**e.** solubility	**g.** solution
b. colloid	**d.** mixture	**f.** solute	**h.** solvent

1. ________ I am smoke, cheese, and foam. I am a mixture that does not settle. Who am I?

2. ________ I am the water in sugar water. Who am I?

3. ________ I am the sugar in sugar water. Who am I?

4. ________ I am steel and I am brass. Who am I?

5. ________ I am the maximum amount of solute that can go into a solvent. What am I?

6. ________ Using evaporation and condensation, I can separate the liquids in a mixture. What am I?

7. ________ I am a combination of two or more materials, but none of my parts are chemically combined. What am I?

8. ________ I can be made with solids, liquids, and gases. All my parts blend so that I look the same everywhere, even under a microscope. Who am I?

Name ______________________ Date ____________

Mixtures

Fill in the blanks.

alloys	distillation	liquids
boiling points	heterogeneous	solids
condensing	homogeneous	suspensions

Several substances that are physically mixed together but not chemically combined are called mixtures. Mixtures can include various combinations of solids, liquids, and gases. Liquids in a mixture may have different ________________ . Boiling and ________________ the liquids, a process called ________________ , can be used to separate them.

There are two kinds of mixtures: those that are the same throughout (________________) and those that are not (________________). Homogeneous mixtures, such as sugar water, are called solutions. Gases form solutions more easily than ________________ do, and liquids form solutions more easily than ________________ do. Solutions of two or more solids are called ________________ .

The different parts of some heterogeneous mixtures can clearly be seen by the naked eye. These are called ________________ . Some suspensions settle to the bottom.

LESSON **Outline**

Name ______________________ Date ____________

Compounds and Chemical Changes

Use your textbook to help you fill in the blanks.

What are compounds?

1. A combination of two or more elements is called a(n) ____________________ .
2. A compound has different properties than do the ____________________ that formed it.
3. Rust is a combination of iron and ____________________ .
4. The chemical name for rust is ____________________ .
5. The chemical formula for rust is ____________________ .

What are chemical changes?

6. Changing one substance into another is a(n) ____________________ .
7. When atoms break their old links and form new links with other atoms, a(n) ____________________ has occurred.
8. Chemists keep track of which substances are used and created in a chemical reaction by writing ____________________ .
9. Chemicals on the left side of a chemical equation are called ____________________ ; chemicals on the right side are called ____________________ .
10. In every chemical reaction, the total mass of the reactants always equals the total mass of the products. This fact is known as the ____________________ .

How can you spot a chemical change?

11. A color change on metal that is caused by a chemical change is called ____________________ .

12. Bubbles form when baking soda and vinegar are mixed, indicating that a ____________________ has taken place.

13. A solid that forms when two solutions are mixed is called a(n) ____________________ .

14. If a chemical reaction produces heat and light, then reversing the reaction should ____________________ .

How can you use chemical changes?

15. Plants use a chemical reaction called ____________________ to produce sugars from sunlight, water, and carbon dioxide.

16. Plants and animals use a chemical reaction called ____________________ to burn sugars for energy.

17. Chemical reactions are used to produce a variety of products, such as ____________________ .

Critical Thinking

18. Write the equation for the chemical change that produces water from two hydrogen molecules and one oxygen molecule. Label the reactants and the products. (Hint: Remember to take into account the conservation of mass.)

____

____

____

Name ____________________ Date __________

Compounds and Chemical Changes

Use the words in the word box to fill in the blanks.

chemical	photosynthesis	reactants
compound	precipitate	tarnish
equations	products	

1. The ________________ are on the left side of a chemical equation.

2. The ________________ are on the right side of a chemical equation.

3. The chemical reaction that plants use to produce sugar is known as ________________ .

4. A solid that is a product of a chemical reaction is called a(n) ________________ .

5. Atoms break their old links and form new links during a(n) ________________ change.

6. Chemists keep track of chemical reactions by using chemical ________________ .

7. A color change in metal caused by a chemical change is called ________________ .

8. A chemical combination of two or more elements is a(n) ________________ .

Name ____________________ Date ____________

LESSON
Cloze Activity

Compounds and Chemical Changes

Fill in the blanks.

chemical equations	$C_6H_{12}O_6$	H_2O
chemical formulas	compounds	left
CO_2	elements	

A chemical change results in one or more products that are different from the reactants. Atoms break their links and form new links with other atoms to form new ____________________ . Chemists describe what goes on in a chemical change by writing ____________________ . The substances to the ____________________ of the arrow in a chemical equation are the reactants; the substances to the right of the arrow are the products. The compounds in a chemical equation are written as ____________________ . A chemical formula tells which ____________________ are in a compound and how many atoms there are of each. For example, the chemical formula for water is ____________________ , and the chemical formula for carbon dioxide is ____________________ . The chemical equation for photosynthesis is $6H_2O + 6CO_2 \xrightarrow{☼}$ ____________________ $+ 6O_2$. The numbers of atoms of each element are the same on each side of the equation.

Name ______________________ Date __________

The Case of the Mystery Compounds

Write About It

Do research and write a report about how scientists can test water for pollutants and dangerous chemical compounds. Which chemical reactions do they use to perform the test? Give the steps of the process in order.

Getting Ideas

As you do research on how scientists test water, fill out the chart below. Write the steps in order.

First

Next

Last

Planning and Organizing

Organize the steps that Sean wrote about testing water for chlorine.

1. Chlorine will turn the litmus paper red, then white. ________
2. Place a sample of the water in a test tube. ________
3. Dip blue litmus paper in the water. ________

Name ______________________ Date ____________

Drafting

Write a sentence to begin your report. Tell an important idea about testing water for pollutants and dangerous chemical compounds.

____

____

____

Now write your report. Use a separate piece of paper. Begin with the sentence you wrote above. Then tell the steps scientists follow to test water. Be sure to include important facts and details about chemical reactions.

Revising and Proofreading

Here are some sentences Sean wrote. They are very wordy. Read each pair. Combine each pair into one sentence by cutting out unnecessary words. Write the new sentence on the line.

1. Make sure the test tube you use is clean. It must be sterile.

   ____

2. The chemical reaction may produce changes in color. It may produce changes in smell.

   ____

3. Test the sample quickly. Do the test within two hours.

   ____

Now revise and proofread your writing. Ask yourself:

- ▶ Did I tell the steps of testing water in order?
- ▶ Did I explain the chemical processes involved?
- ▶ Did I correct all errors?

Name ______________________ Date ____________

Comparing Kinds of Matter

Choose the letter of the best answer.

1. A material that cannot be broken down into simpler chemical substances is a(n)

a. element.

b. metal.

c. chemical.

d. molecule.

2. The positively charged particles in an atom are called

a. neutrons.

b. electrons.

c. protons.

d. molecules.

3. Which particles in an atom are negatively charged?

a. protons

b. neutrons

c. molecules

d. electrons

4. Two or more atoms can join to form a(n)

a. element.

b. neutron.

c. molecule.

d. superatom.

5. The amount of matter in an object is its

a. weight.

b. mass.

c. volume.

d. density.

6. The strength of gravity on an object determines that object's

a. mass.

b. volume.

c. weight.

d. electrical charge.

7. The amount of space being taken up by matter is known as its

a. volume.

b. weight.

c. mass.

d. density.

8. The amount of mass for each milliliter of a substance determines the substance's

a. weight.

b. buoyancy.

c. density.

d. volume.

Name ______________________ Date ____________

Choose the letter of the best answer.

9. Which of the following is a physical change?

a. paper burning

b. egg frying

c. water boiling

d. baking soda and vinegar fizzing

10. Snow changing to water vapor is an example of

a. sublimation.

b. boiling.

c. melting.

d. thermal contraction.

11. When a gas loses heat, it

a. evaporates.

b. freezes.

c. sublimates.

d. condenses.

12. The temperature at which alcohol changes to a gas is its

a. sublimation point.

b. freezing point.

c. boiling point.

d. melting point.

13. Steel is an example of a(n)

a. alloy.

b. colloid.

c. heterogeneous mixture.

d. suspension.

14. In a saltwater solution, the salt is a(n)

a. alloy.

b. colloid.

c. solvent.

d. solute.

15. Which of the following is a compound?

a. brass

b. rust

c. iron

d. steel

Name ______________________ Date ____________

Electricity and Magnetism

Complete the concept map about electricity and magnetism. Some parts have been done for you.

Electricity

Nature and Movement	Control of Movement	Uses
result of electrical charge; can be ______________ similar charges ______________; ______________ charges attract; charged particles ______________ in electric current	made easier by ______________; ______________ by insulators; carried along circuit; made of power source, ______________, and load	converted into heat, light, or ______________

Magnetism

Nature	Electromagnets	Producing Electricity
each magnet has ______________ poles; similar poles ______________; opposite poles ______________; stength represented by ______________ lines	moving ______________ create magnetic fields; the stronger the ______________ in the wire, the stronger the field	A ______________ uses a moving coil in a magnetic field to produce an electric current.

Name ______________________ Date ____________

LESSON **Outline**

Static Electricity

Use your textbook to help you fill in the blanks.

What is electrical charge?

1. Electrical charge is a(n) ____________________ .
2. Scientists call the two types of electrical charges ____________________ and ____________________ .
3. When positive and negative charges ____________________ , the matter is said to be neutral.
4. Like charges repel or ____________________ each other, but opposite charges ____________________ , or pull toward each other.
5. When two objects touch, ____________________ move between them.
6. Negative charges move more ____________________ than positive charges.

What is static electricity?

7. The buildup of electrical charges on an object is called ____________________ .
8. Rubbing one object against another causes a ____________________ of one kind of charge.
9. When you rub a balloon on wool, negative charges build up in the ____________________ .

Name ______________________ Date __________

10. When you place the balloon against a wall, it ______________ positive charges in the wall. Because of that, the balloon ______________ against the wall.

What is an electrical discharge?

11. Lightning is the discharge of ______________ inside a storm cloud.

12. A(n) ______________ is the movement of static electricity from one object to another.

13. When lightning strikes, ______________ in a cloud push down on the negative charges in the ground.

14. The safest place in a lightning storm is ______________ .

What are conductors and insulators?

15. Copper and silver are good ______________ because charges flow through them easily.

16. The outside of an electrical wire is covered by a(n) ______________ such as rubber or plastic.

17. The insulator keeps the electricity inside the wire and ______________ .

Critical Thinking

18. Suppose you walked on a carpet and built up a charge of static electricity. Would you feel a shock if you touched a plastic cup? Why or why not?

____

____

____

Name ______________________ Date __________

Static Electricity

Match the correct word or words to their definitions by writing the letter in the space provided.

a. attract	**d.** electrical charge	**g.** repel
b. conductors	**e.** insulators	**h.** static electricity
c. discharge	**f.** neutral	

1. ________ Rubber, plastic, and glass are good examples of these materials.

2. ________ Two objects that both have negative charges will do this to each other.

3. ________ This property of matter has two types, positive and negative.

4. ________ When clothes stick together after coming out of a clothes dryer, they might have this buildup.

5. ________ Copper and other metals are good examples of these materials.

6. ________ Walking across a carpet and then touching something metal can cause this movement of electricity.

7. ________ Objects with a negative charge will do this to objects with a positive charge.

8. ________ Objects that have an equal number of positive and negative particles are said to be this.

Name ______________________ Date ____________

Static Electricity

Use the words in the box to fill in the blanks.

charged particles	insulators	static electricity
conductors	negative	
electrical charges	positive	

Electricity powers traffic lights, appliances, and computers. There are different kinds of electricity, but all electricity is the result of ____________________ . There are two types of electrical charges. Scientists call these charges ____________________ and ____________________ . When two objects touch, ____________________ can move from one object to the other.

The buildup of electrical charges on an object is called ____________________ . It is what makes clothes stick together.

Metals such as copper and silver are examples of good ____________________ because they let charges flow through them easily. Rubber, plastic, and glass are examples of good ____________________ . These materials do not let charges flow through them easily.

Name ______________________ Date ____________

LESSON **Outline**

Current Electricity

Use your textbook to help you fill in the blanks.

How can electricity flow?

1. Circuits must have an unbroken path of conductors and a(n) ____________________ that causes the electrons to move along the path.

2. A device that can open or close a circuit is called a(n) ____________________.

3. Electrons in a circuit are pushed away from a battery's ____________________ pole and are pulled toward the battery's ____________________ pole.

4. Resistance is measured in ____________________, and electric current is measured in ____________________.

5. A current of electrons moving through resistors loses energy that changes into ____________________.

What kinds of circuits are there?

6. A circuit with only one conductive path is a(n) ____________________ circuit; a circuit with more than one conductive path is a(n) ____________________ circuit.

Name ______________________ Date ____________

7. In a series circuit, resistance increases with each ______________ added.

8. In a parallel circuit, paths with greater ______________ have less electric current flowing through them.

How can you use electricity safely?

9. To protect against large currents, homes have ______________ .

10. Outlets in kitchens and bathrooms have ______________ that turn an outlet off if a short is detected.

11. Touching two ______________ at the same time or touching one power line and ______________ or some grounded object can be deadly.

Critical Thinking

12. When a home circuit breaker opens, the lights in some rooms go off but the lights in other rooms stay on. Explain why.

____

____

____

Name ______________________________ Date ____________

Current Electricity

Matching

Match the correct letter with the description.

a. circuit	**d.** resistor
b. electric current	**e.** switch
c. fuse	**f.** voltage source

1. ________ a wire that breaks if too much current flows through it

2. ________ device that opens or closes an electric circuit

3. ________ an unbroken path of conductors through which electric current passes

4. ________ a device to move electrons along a path, often a battery

5. ________ an object in an electric circuit that resists the flow of electrons

6. ________ a flow of electricity through a conductor

LESSON
Cloze Activity

Name ______________________________ Date ______________

Current Electricity

Fill in the blanks.

circuit	negative	static electricity
conductor	poles	switch
electric current	resistor	

When objects rub against each other, electrons sometimes move from one object onto the other. The resulting buildup of charged particles is called __________________. Electricity moves more easily through a wire. When a wire is hooked to the two oppositely charged __________________ of a battery, a(n) __________________ is formed. Electrons move from the __________________ pole through the __________________ to the positive pole. This flow of electrons is called a(n) __________________.

A device that opens or closes a circuit is called a(n) __________________. Any device, such as a light bulb, that resists the flow of electrons is a(n) __________________. Circuits that have only one path for electrons are series circuits, and circuits that have more than one path are parallel circuits.

Name ______________________ Date ____________

LESSON **Outline**

Magnetism

Use your textbook to help you fill in the blanks.

What is magnetism?

1. When a magnet is cut in half, each of the two pieces has a(n) ______________ pole and a(n) ______________ pole.
2. Like poles of a magnet ______________ each other, and unlike poles ______________ each other.
3. The Earth is a giant permanent ______________ .
4. Whenever an electric charge moves, it creates ______________ forces.
5. The ______________ together the lines of a magnetic field, the stronger the magnetic force.

What are electromagnets?

6. An electric current that produces a magnetic field is called a(n) ______________ .
7. A magnetic field goes in ______________ around a straight wire when current is flowing through it.
8. Wrapping many loops of wire together ______________ the magnetism of the coil.
9. You can increase the strength of an electromagnet in three ways: ______________ , place an iron rod inside the coils, or ______________ .

Name ______________________________ Date ____________

10. As the electric current rises and falls in the ____________ of a speaker, its magnetic field changes, causing a cone of paper or metal to vibrate.

11. In an electric motor, a coil acting as an electromagnet rotates between the poles of a(n) ____________.

How can magnets produce electricity?

12. A generator creates an electric current by spinning a coil of wire between the poles of a powerful ____________.

13. The energy needed to spin the coils in an electric generator can come from ____________ in a hydroelectric dam, ____________ in a coal-fired power plant, or from wind or tides.

What is magnetic levitation?

14. Two electromagnets can push against each other to ____________ an object.

15. Scientists have designed ____________ trains that are held just above their tracks by electromagnets, greatly reducing the amount of energy lost to ____________.

Critical Thinking

16. In what way is an electric generator the opposite of an electric motor?

____

____

____

____

Name ______________________________ Date ____________

Magnetism

Who am I? What am I?

Choose a word from the word box below that answers each question.

a. alternating current	**d.** magnetic field
b. electromagnet	**e.** magnetic levitation
c. generator	**f.** magnetism

1. ________ When my wire coils are spun between the poles of my powerful magnet, I produce electricity. Who am I?

2. ________ I move back and forth through a wire, changing directions many times per second. What am I?

3. ________ I use magnetic forces to lift objects. I can even lift an entire train! Who am I?

4. ________ I am magnetic when an electric current flows through me. Who am I?

5. ________ When you have me, you can push or pull on another object that also has me. What am I?

6. ________ I describe the strength and direction of a magnet's force. If you sprinkle iron filings near a magnet, you can see me. What am I?

LESSON
Cloze Activity

Name ______________________________ Date ____________

Magnetism

Fill in the blanks.

electric current	magnetic field	poles
electric motor	north	south
electromagnet	permanent magnet	spin

Permanent magnets are made of metals such as iron. They have two ________________, north and south, and a(n) ________________ around them. An iron core with a wire coil wrapped around it is called a(n) ________________. When a(n) ________________ passes through the wire coil, a magnetic field with a(n) ________________ and a(n) ________________ pole is generated.

Electric motors and electric generators have an electromagnet between the poles of a very strong ________________. In a(n) ________________, current is sent through the wire coil. The poles of the electromagnet switch back and forth, causing it to ________________ between poles of the permanent magnet. In an electric generator, energy from falling water or some other source is used to spin the wire coil past the poles of the permanent magnet, generating electricity in the wire coil.

Name ______________________ Date ______________

Reading in Science

Building a Better Battery

Read the Reading in Science feature in your textbook. Try to draw conclusions from text clues.

Draw Conclusions

Fill in the Drawing Conclusions Chart using text clues you find in the article.

Text Clues	Conclusion
1. Batteries are devices that store ______________ and make it available in ______________ form.	Batteries convert ______________ energy to ______________ energy.
2. All batteries have positive and negative ______________ and an electrolyte through which a(n) ______________ can flow.	Electrodes are ______________ to which wires can be attached; the ______________ is a solution through which electrons move.
3. A(n) ______________ battery has two plates in a(n) ______________ solution; most cars have them.	Lead plates are the ______________ of this type of battery, and ______________ solution is the electrolyte; this type of battery recharges when the car runs.
4. Hybrid cars use ______________ batteries that recharge when the car slows down.	Nickel hydride batteries are used in ______________ cars, instead of lead-acid batteries.

Name ______________________________ Date ____________

Write About It

Draw Conclusions Explain what is used as an electrolyte in a voltaic pile. Explain what types of batteries are used in hybrid cars.

Planning and Organizing

Answer these questions in more detail.

Describe the components of a battery.

____

____

Explain what an electrolyte is.

____

Explain how a voltaic pile is constructed and what is used as the electrolyte.

____

____

What kinds of batteries do cars have and why do they have them?

____

____

What type of batteries are used in hybrid cars? Why?

____

____

Name ______________________ Date ____________

Electricity and Magnetism

Choose the letter of the best answer.

1. A flow of electric charge through a material is called

a. current electricity

b. static electricity

c. voltage

d. insulation

2. The sudden movement of static electricity is called a

a. circuit

b. discharge

c. electromagnet

d. conductor

3. A material that allows electric charges to flow through it easily is called

a. an insulator

b. a conductor

c. a resistor

d. a switch

4. Which of these is made up of parts that work together to allow electricity to flow?

a. circuit

b. insulator

c. current

d. electromagnet

5. When a charged object is placed near a neutral object, the result is

a. an induced charge

b. a magnetic field

c. a parallel circuit

d. an electromagnet

6. A device that is magnetized by current electricity is called

a. a generator

b. an electromagnet

c. a series circuit

d. a parallel circuit

7. A device that indicates the direction of Earth's magnetic North Pole is a

a. generator

b. motor

c. compass

d. sonar

Name ____________________ Date __________

8. Which of these has magnetic properties and can be controlled by a switch?
   - **a.** an electric current
   - **b.** an electromagnet
   - **c.** a magnetic field
   - **d.** a magnetic force

9. Which of these is produced by the buildup of negative charges on a material?
   - **a.** conduction
   - **b.** magnetism
   - **c.** static electricity
   - **d.** current electricity

10. The region of magnetic force around a magnet is called a
   - **a.** magnetic field
   - **b.** compass
   - **c.** generator
   - **d.** electromagnet

11. Which of the following is a resistor?
   - **a.** a light bulb
   - **b.** a switch
   - **c.** a battery
   - **d.** a wire

12. What happens when one bulb in a series circuit burns out?
   - **a.** The other bulbs get dimmer.
   - **b.** The other bulbs go out.
   - **c.** The other bulbs get brighter.
   - **d.** The other bulbs stay the same.

13. Which statement about magnets is true?
   - **a.** Like poles attract each other.
   - **b.** Unlike poles attract each other.
   - **c.** Unlike poles repel each other.
   - **d.** Poles have no affect on each other.

14. A device that changes electrical energy into a spinning motion is a(n)
   - **a.** electric motor.
   - **b.** transformer.
   - **c.** electromagnet.
   - **d.** electric generator.

Name ______________________ Date ____________

Green and Clean: Plants as Pollution Control

Read the Literature feature in your textbook.

Write About It

Response to Literature This article describes how plants are used to help clean polluted soil. Research additional information about cleaning up waste. Write a report about the cleaning process. Include facts and details from this article and from your research.